68th Porcelain Enamel Institute Technical Forum

68th Porcelain Enamel Institute Technical Forum

A collection of papers presented at the 68th Porcelain Enamel Institute Technical Forum, May 15–18, 2006, Nashville, Tennessee

Conference Director
Holger Evele

Assistant Conference Director
Peter Vodak

Editor
William D. Faust

The American Ceramic Society

BICENTENNIAL
1807
WILEY
2007
BICENTENNIAL

A JOHN WILEY & SONS, INC., PUBLICATION

Published by John Wiley & Sons, Inc., Hoboken, New Jersey
Published simultaneously in Canada.

For general information on our other products and services please contact our Customer Care
Department within the U.S. at 877-762-2974, outside the U.S. at 317-572-3993 or fax 317-572-4002.

Wiley also publishes its books in a variety of electronic formats. Some content that appears in print,
however, may not be available in electronic format.

Library of Congress Cataloging-in-Publication Data is available.

ISBN-13 978-0-470-09735-9
ISBN-10 0-470-09735-3

Printed in the United States of America.

10 9 8 7 6 5 4 3 2 1

Contents

Wet Porcelain Enamels and Processing

Preface

The entire Technical Forum Committee is pleased to deliver to you the proceedings of the 68th Annual PEI Technical Forum. This volume represents the successful completion of a year's worth of planning and preparation, culminating in three days of meetings and seminars at the Double Tree Downtown Hotel in Nashville, Tennessee on May 15–18, 2006. As you receive these proceedings, work is already progressing on the 69th Technical Forum, to be held September 18–20, 2007 in Indianapolis, Indiana in conjunction with the FIN-X '07 International Expo & Conference for Industrial Finishers.

I would like to thank my vice-chairman, Peter Vodak (Engineered Storage Products Company) as well as members of the Technical Forum Committee for their time, efforts, and supportive endeavors on behalf of this year's forum. The success of the forum is directly attributable to their contributions. Each year we strive to uphold the tradition of offering information that has both useful and practical applications for our industry. I believe we have accomplished that challenge.

The Back-To-Basics Seminar continues to be an important element of the Technical Forum. Attendance at the B2B continues to be consistently good. Our appreciation goes to the faculty staff for another very outstanding program. This seminar continues to be a well-attended favorite, attracting both newcomers to the porcelain enamel industry, as well as seasoned veterans. Again, thanks to all involved.

The thanks of the entire Committee goes out to this year's excellent group of speakers who provided us with information on the latest in materials and equipment used in the porcelain enameling process. We are grateful to them for their time and efforts in researching, preparing and presenting their informative papers. Further thanks go to those suppliers who participated and supported the ever-popular Supplier's Mart. Our final thanks go to those who attended and participated in the 68th Annual PEI Technical Forum.

Peter Vodak (Vice Chairman) and I will again head up the planning efforts for the 2006 Technical Forum. We look forward to the 2007 PEI Technical Forum in Indianapolis which promises to be informative and worthwhile.

HOLGER F. EVELE
Ferro Corporation
Chairman 2006 PEI Technical Forum Committee and Back-To-Basics Seminar

2006 PEI Officers

Chairman of the Board
JACK MCMAHON
Pemco Corporation

President
BOB HARRIS
Hanson Industries

Vice Presidents
BILL GANZER
Mapes & Sprowl Steel

KEN KREEGER
Nordson Company

DON MCCORMICK
Electrolux Home Products

TIM SCOTT
Henkel Surface Technologies

NICK SEDELIA
Whirlpool Corporation

MILES VOLTAVA
Ferro Corporation

PAT WALSH
Porcelain Industries

2006 Technical Forum Committee

Chairman: Holger Evele, Ferro Corporation
Vice Chairman: Peter Vodak, Engineered Storage Products

Peter Dority, Coral Chemical Company
Dough Giese, GE Appliances
Cullen Hackler, Porcelain Enamel Institute
Mike Horton, KMI Systems
Ken Kreeger, Nordson Corporation
Erik Miller, Maytag Cooking Products
Liam O'Byrne, O'Byrne Consulting Services
Tim Scott, Henkel Surface Technologies
Lester Smith, Porcelain Consultants
Larry Steele, Mapes & Sprowl Steel
Dave Thomas, American Trim
Peter Vodak, Engineered Storage Products
Miles Votava, Ferro Corporation
Jack Waggener, URS Corporation
Mike Wilczynski, AO Smith Protective Coatings
Ted Wolowicz, Electrolux Home Products
Jeff Wright, Ferro Corporation

Porcelain Enamel Institue
PO Box 920220
Norcross, GA 30010
Phone : 770-281-8980
E-mail : penamel@aol.com
www.buyporcelain.org
www.porcelainenamel.com

Past PEI Technical Forum Committee Chairs

Steve Kilczewski   2004–05
Pemco Corporation

Liam O'Byrne   2002–03
AB&I Foundry

Jeff Sellins   2000–01
Maytag Cooking Products

Robert Reese   1998–99
Frigidaire Home Products

David Thomas   1996–97
The Erie Ceramic Arts Company

Rusty Rarey   1994–95
LTV Steel Company

Douglas Giese   1992–93
GE Appliances

Anthony Mazzuca   1990–91
Mobay Corporation

William McClure   1988–89
Magic Chef

Larry Steele   1986–87
Armco Steel

Donald Sauder   1984–85
WCI-Range Division

James Quigley   1982–83
Ferro Corporation

George Hughes   1980–81
Vitreous Steel Products Company

Lester Smith   1978–79
Porcelain Metals Corporation

A. I. Andrews
Memorial Lecture

NANOTECHNOLOGY – CHALLENGES AND OPPORTUNITIES FOR THE FUTURE

Richard A. Haber
Ceramic and Composites Material Center
Rutgers University, New Jersey

Abstract
Nanotechnology encompasses a broad range of materials and processes on a sub-micron scale. Nanotechnology materials have been in use for many years. However, recent developments in synthesis by high temperature chemistry have yielded new material structures. These new material exhibit new and unexpected properties and behaviors.

Adaptation of these structures is the key to taking advantage of these materials. Current nanotechnology applications cover areas of rheological control (aerogels), composite structures (grapheme sheets) and surface modification (transparent hydrophobic coatings). Future uses of nano-materials are envisioned in information storage in the gigabyte range, antibacterial and anti-fungal agents with surface active components, detoxification of contaminated clothing, fuel cells, catalyst with specific structures, energy storage of exceptionally high density such as hydrogen storage, and medical applications such as targeted drug delivery vehicles for tumors and viruses.

The growth of nano-materials in many products is indicated to have impacts as significant as antibiotics, printed circuits and manmade polymer.

Outline of Presentation

- **What are Nanomaterials?**
- **Why Are We Excited?**
- **A Perspective on Nanopowders in 2006**

GROWTH
U.S. nanomaterials markets to expand significantly

$ MILLIONS	2002	2007	2012	2020	ANNUAL GROWTH 2002–20
Minerals	$160	$675	$2,100	$11,500	28%
Metals	45	150	500	3,000	26
Polymers & chemicals	5	175	1,400	15,500	56
New materials[a]	10	100	500	5,000	41
TOTAL	$200	$1,100	$4,500	$35,000	33%

a Includes carbon nanotubes. **SOURCE:** Freedonia Group

APPLICATIONS
Nanomaterials is focus of most nanotech start-ups

- Other[a] 16%
- Consumer products 7%
- Electronics 11%
- Research 14%
- Materials & production processes 31%
- Medicine/ pharmaceuticals 21%

U.S. start-up companies = 150

NOTE: Data as of late 2003. a Includes telecommunications, analysis, information technology, and energy storage at 4% each. **SOURCE:** NanoBusiness Alliance

THE STATE UNIVERSITY OF NEW JERSEY
RUTGERS

Ceramic and Composite Materials Center

What are Nanomaterials?

Structures or materials who have characteristic length scales (sizes) in 1-D, 2-D or 3-D that are ~ 10^{-9} meters (nano-meters, nm, 1/1,000,000,000 of a meter, 1 nm is ~ 3-4 atoms wide)

Materials, structures or systems where behavior and control of processes are at the atomic or molecular level

Materials that exhibit novel (unexpected) properties, behavior and phenomena

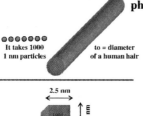

It takes 1000
1 nm particles   to = diameter
of a human hair

2.5 nm

2.5 nm

A box 2.5 nm on a side would
hold ~ 1000 atoms

- The Rutgers University budget is ~ $ 1,000,000,000. Thus $1.00 is a nano-element of the budget

- 4 mm is a nano-element of the distance from NY to LA

Ceramic and Composite Materials Center

Nanotechnology : The Challenge

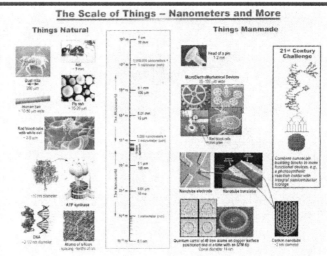

The Scale of Things -- Nanometers and More

Lucent Technologies
Bell Labs Innovations

agere systems

Nano Enamel...

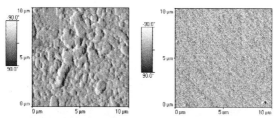

AFM images of pellicle-covered enamel (left, below) and sound enamel before covering (right).

Ceramic and Composite Materials Center

Nano-electronics, quantum computing, super-capacitors, super-batteries, bio-hazard sensors, 1000 CD's on your wrist, novel drugs and drug delivery systems, cellular repair, artificial organs, miniature airplanes, e-ink, self assembly, super-armor, transparent armor, lab on a chip, nano- bio-informatics

Total societal impact far greater than silicon microelectronics integrated circuit revolution of the 20th century

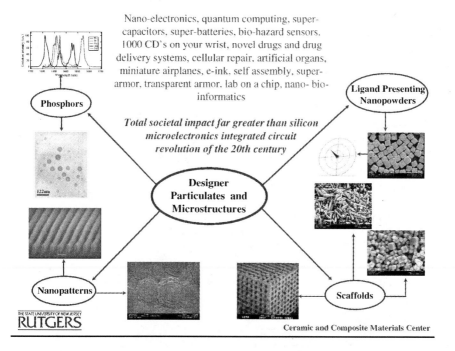

Phosphors

Ligand Presenting Nanopowders

Designer Particulates and Microstructures

Nanopatterns

Scaffolds

Ceramic and Composite Materials Center

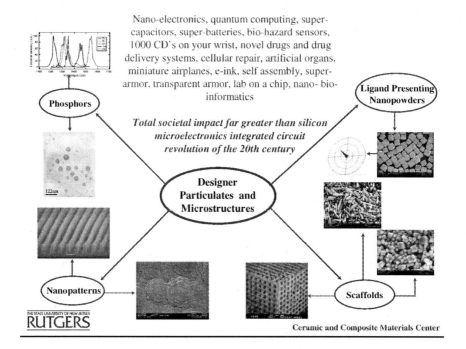

Nano-electronics, quantum computing, super-capacitors, super-batteries, bio-hazard sensors, 1000 CD's on your wrist, novel drugs and drug delivery systems, cellular repair, artificial organs, miniature airplanes, e-ink, self assembly, super-armor, transparent armor, lab on a chip, nano- bio-informatics

Total societal impact far greater than silicon microelectronics integrated circuit revolution of the 20th century

Phosphors

Ligand Presenting Nanopowders

Designer Particulates and Microstructures

Nanopatterns

Scaffolds

RUTGERS THE STATE UNIVERSITY OF NEW JERSEY

Ceramic and Composite Materials Center

Production of Pyrogenic Metal Oxides –

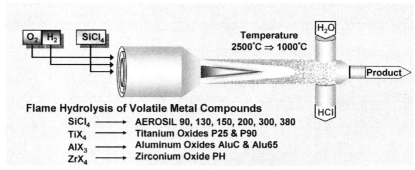

Temperature
2500°C ⇒ 1000°C

O₂ H₂ SiCl₄

H₂O

Product

HCl

Flame Hydrolysis of Volatile Metal Compounds

$SiCl_4 \longrightarrow$ AEROSIL 90, 130, 150, 200, 300, 380
$TiX_4 \longrightarrow$ Titanium Oxides P25 & P90
$AlX_3 \longrightarrow$ Aluminum Oxides AluC & Alu65
$ZrX_4 \longrightarrow$ Zirconium Oxide PH

(Process invented 1941 by Harry Kloepfer, named AEROSIL® in '43 when commercial production began – '53 Al_2O_3, '54 TiO_2, '63 Hydrophobic grades, '64 AERODISP®)

RUTGERS THE STATE UNIVERSITY OF NEW JERSEY

Ceramic and Composite Materials Center

Fundamentals of Particle Formation

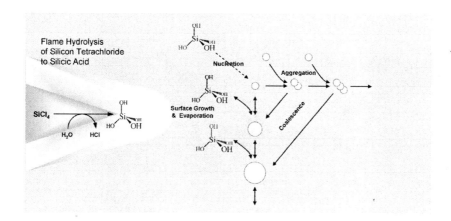

Ceramic and Composite Materials Center

Transmission Microscope (TEM) Pictures

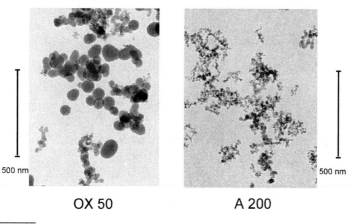

OX 50   A 200

Ceramic and Composite Materials Center

Flame Hydrolysis: Nanoscale Oxides

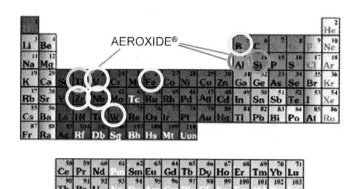

Ceramic and Composite Materials Center

Where is Nanotechnology being commercialized today?

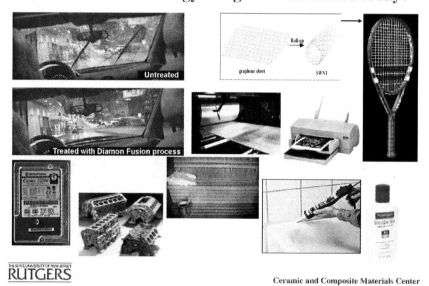

Ceramic and Composite Materials Center

Transparent UV Absorption & Conductivity

• UV-Absorption: Nano structured cores based on ZnO and TiO_2

• ZnO based core effective UV-A filter, absorbs and reflects UV-light in comparison to organic UV filters offers high temperature stability and no leaching.

• Primary Core size is 25 nm with aggregates range from 80 – 200nm

• Small primary core size gives rise to improved transparency and durability over organic UV-filters

• Surface treatment (TMOS) to improve hydrophobicity and optimize dispersibility in selected coating formulations

• Conductivity: Nano structure core based on In_2O_3 / SnO_2 (Indium- Tin oxide)

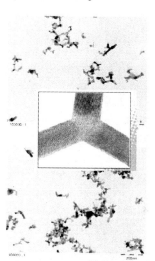

RUTGERS
THE STATE UNIVERSITY OF NEW JERSEY

Ceramic and Composite Materials Center

Sunblock Characteristic UV Spectra

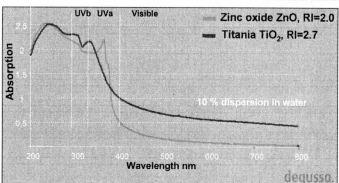

RUTGERS
THE STATE UNIVERSITY OF NEW JERSEY

Ceramic and Composite Materials Center

Nanomaterials Over 30 Years!

'83: Apple PC, No Hard Drive 0 Bytes

'03: 200 GB Hard Drives

'13: 10 TB Hard Drives

10,000,000,000,000 Bytes

Imagine a Palm Pilot that could readily contain the information and knowledge of every volume in the Library of Congress!

THE STATE UNIVERSITY OF NEW JERSEY
RUTGERS

Ceramic and Composite Materials Center

Nanomaterials for Information Storage

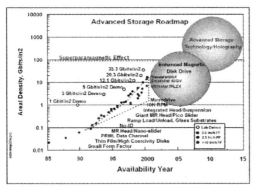

- **Self Assembled/ Ordering of 4 nm Fe/Pt nanoparticles**
- **Nanomaterials will yield areal storage densities > 100 Gigabits/square inch**

T. Tsakalakos, Rutgers University

THE STATE UNIVERSITY OF NEW JERSEY
RUTGERS

Ceramic and Composite Materials Center

Nanomaterials as Effective Bactericidal and Fungicidal Agent

This washing machine from Samsung uses silver to kill germs on clothes.

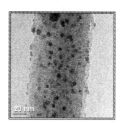

nano-silver colloids and silver nanoparticles in different natural templates

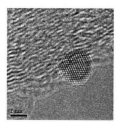

multimetallic nano colloids – 75% Cu and 25% Au

THE STATE UNIVERSITY OF NEW JERSEY
RUTGERS

From MJ Yucaman,2005

Ceramic and Composite Materials Center

Reactive Catalysts in Fabrics that Detoxify Contaminants In Clothing within Hours of Exposure

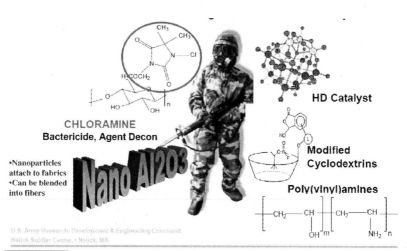

CHLORAMINE
Bactericide, Agent Decon

•Nanoparticles attach to fabrics
•Can be blended into fibers

HD Catalyst

Modified Cyclodextrins

Poly(vinyl)amines

U.S. Army Research, Development & Engineering Command
Natick Soldier Center, • Natick, MA

THE STATE UNIVERSITY OF NEW JERSEY
RUTGERS

H. Schreuder-Gibson

Ceramic and Composite Materials Center

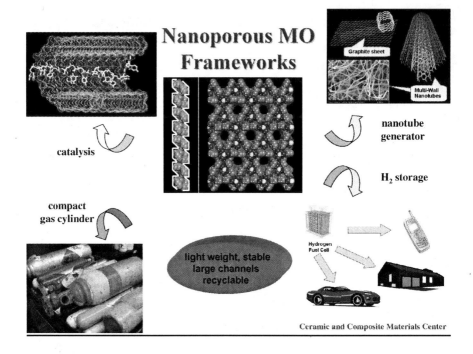

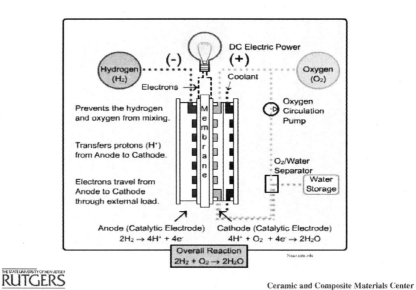

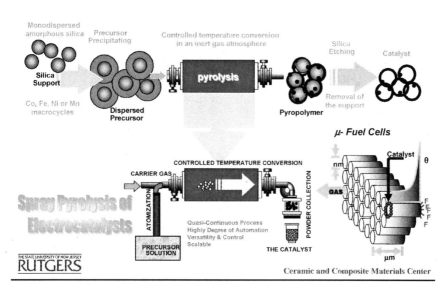

Non-Platinum, Templated Electrocatalysts For Fuel Cell Applications

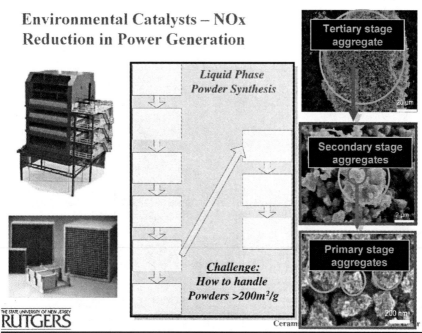

Environmental Catalysts – NOx Reduction in Power Generation

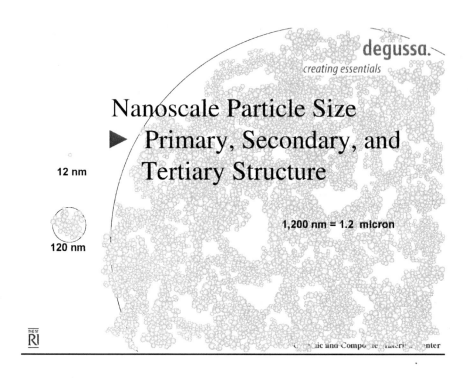

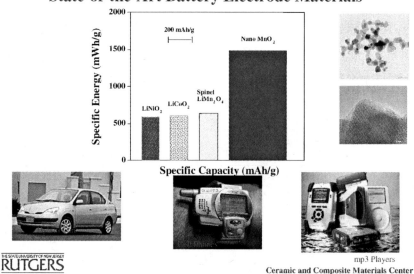

Optical Nanocomposites for Planar Waveguides

- •*Nanometer sized particles are below Rayleigh scattering limit*
- •*Nanometer scale homogeneity*
- •*Rare Earth doping of alumino-silicate nanopowders above solubility limit*
- •*CF-CVC production of nanopowders*
- • *Fabricate composite planar waveguide with unique optical properties*

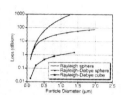

Fig. 1. Rayleigh and Rayleigh–Debye scattering for 10% particle loading and $\Delta n = 0.015$

Cold pressed compact

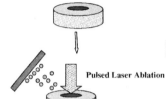

Pulsed Laser Ablation

Lossless Optical Planar Splitter

G. Sigel

Revolutionary Telecomunications Devices From Nanomaterials

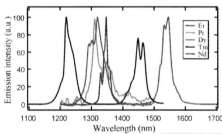

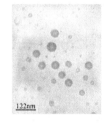

122nm

R.E. Riman, Rutgers University

Mixed Pyrogenic Oxide Capabilities

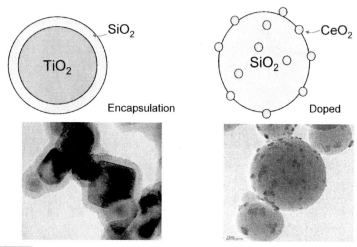

THE STATE UNIVERSITY OF NEW JERSEY
RUTGERS

Ceramic and Composite Materials Center

Patterning the Catalyst/CNT

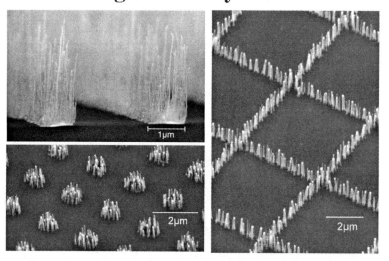

THE STATE UNIVERSITY OF NEW JERSEY
RUTGERS

Manish Chhowalla, Rutgers University
Ceramic and Composite Materials Center

ZnO Nanostructures and Their Applications

- ZnO has wide and direct bandgap of 3.3 eV which can be engineered by alloying with MgO or CdO ; it is piezoelectric.
- ZnO nanotips and nanotip arrays have applications in:
  - field-emission (displays and near field optical probing)
  - nano-lasers
  - photonic bandgap devices (PBG)
  - biomedical sensors
    (DNA, protein binding sites, etc.)
  - STM & AFM

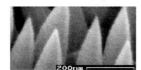

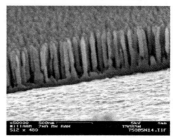

ZnO nanotips on C-sapphire   ZnO nanotip array on GaN

RUTGERS
THE STATE UNIVERSITY OF NEW JERSEY

Y. Lu Electrical and Computer Engineering
Supported by NSF Grant CR -0103096

Ceramic and Composite Materials Center

Carbon Nanotubes (CNT's) for Flat Panel Displays

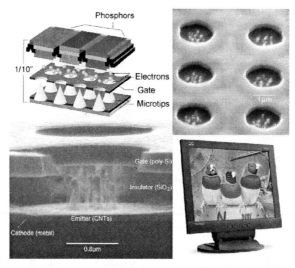

Phosphors

1/10"

Electrons
Gate
Microtips

Gate (poly-Si)
Insulator (SiO₂)
Emitter (CNTs)
Cathode (metal)
0.8µm

RUTGERS
THE STATE UNIVERSITY OF NEW JERSEY

Ceramic and Composite Materials Center

Ceramic Armor Materials

- Superhard materials with effective plasticity
- Transparent materials

Ceramic and Composite Materials Center

Plasma Melting Process for New Metastable Materials

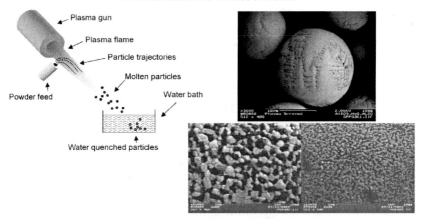

Al_2O_3 – 40 vol. % $MgAl_2O_4$ Plasma Melted Powders

Ceramic and Composite Materials Center

Plasma Melting Process for New Metastable Materials

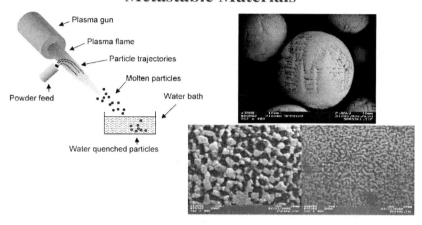

Al$_2$O$_3$ – 40 vol. % MgAl$_2$O$_4$ Plasma Melted Powders

THE STATE UNIVERSITY OF NEW JERSEY
RUTGERS

Ceramic and Composite Materials Center

Nanoparticle Applications A Process Aids

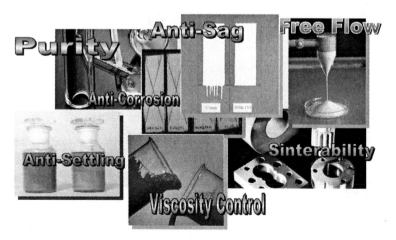

THE STATE UNIVERSITY OF NEW JERSEY
RUTGERS

Ceramic and Composite Materials Center

Major Functions of Fumed Nanoparticles

- **Viscosity & Rheological Control**
    - •Hydrophilic & Hydrophobic
- **Improves Pigment Suspension**
    - •Hydrophilic & Hydrophobic
- **Anti-Corrosion / Water Repellency**
    - •Hydrophobic only
- **Reinforcement of Silicone Rubber/Sealants**
    - •Hydrophilic & Hydrophobic
- **Improves FreeFlow of Powder Formulations**
    - •Hydrophilic & Hydrophobic

THE STATE UNIVERSITY OF NEW JERSEY
RUTGERS

Ceramic and Composite Materials Center

Nanoparticulate control of viscosity in liquids through gel formation

- **reversible formation of a three-dimensional gel structure**

- **especially efficient in liquids of low polarity, e.g. cosmetic oils**

- **viscosity increasing agent (pseudoplastic / thixotropic behavior)**

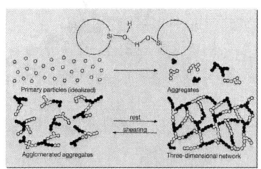

THE STATE UNIVERSITY OF NEW JERSEY
RUTGERS

Ceramic and Composite Materials Center

Interaction of Silica Particles – Thixotropy

3 - Dimensional Network

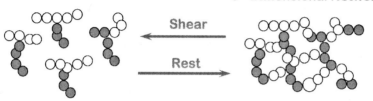

Ceramic and Composite Materials Center

Rheology Control in Coatings

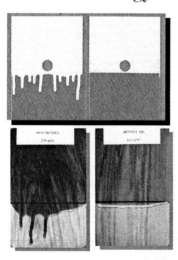

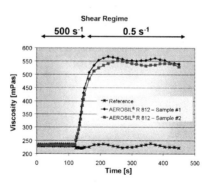

Ceramic and Composite Materials Center

Clear Latex Sealant

Aerosil 200

Aerodisp W7520

THE STATE UNIVERSITY OF NEW JERSEY
RUTGERS

Ceramic and Composite Materials Center

Nanosilica: its mechanism of action

- **The action of nanoparticulate silica is based essentially on three effects:**

  – **gel formation**

  – **coating of solid particles**

  – **adsorption**

THE STATE UNIVERSITY OF NEW JERSEY
RUTGERS

Ceramic and Composite Materials Center

Variations of nanoparticulate oxides

$(CH_3)_3$ Si   Si$(CH_3)_3$

Nanoscale particles, untreated
▸ SiO_2
▸ Al_2O_3
▸ TiO_2
▸ Mixed Oxides

Nanoscale particles, surface treated with
▸ Silanes
▸ Siloxanes
▸ Organic Molecules

Nanoscale particles, structurally modified by
▸ Granulation process
▸ Mechanical treatment

THE STATE UNIVERSITY OF NEW JERSEY
RUTGERS

Ceramic and Composite Materials Center

Optimized Dispersion Improves Efficiency

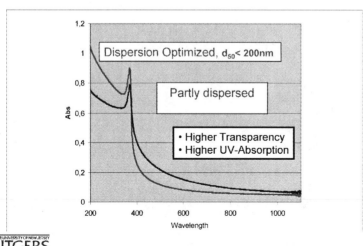

THE STATE UNIVERSITY OF NEW JERSEY
RUTGERS

Ceramic and Composite Materials Center

Nanosilica improves powder flow by coating solid particles

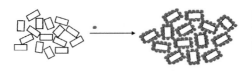

- reduces attraction between particles
- "ball bearing effect"
- smoothes the surface of particles
- prevents incompatibilities between substances
- anti-caking agent, free flow agent, suspending agent

RUTGERS

Ceramic and Composite Materials Center

Coating of solid particles

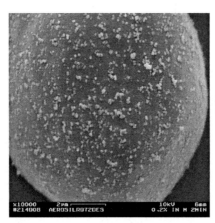

Cornstarch (Cerestar ®)
w/ 0.5 wt.% AEROSIL® R 972

RUTGERS

Ceramic and Composite Materials Center

Frictionizing – Anti-Slip, Anti-Wear

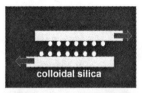

Effect

- **Fumed particles enhance the friction between surfaces, e.g. paper, particles, fiber, etc.**
- **Application by rolling or spraying – easy to Clean!**
- **Much more effective than colloidal silica (usually used at ~20-30%), we recommend ~5% fumed dispersion**

RUTGERS

Ceramic and Composite Materials Center

SEM of Isolated Fumed Silica Particle

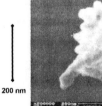

200 nm

(Chem. Eng. Technol. **21** (1998), 745-752)

RUTGERS

Ceramic and Composite Materials Center

Surface Modification of Powder

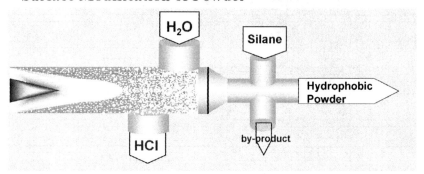

Hydrophobizing Treatments Include:

Dimethyldichlorosilane (DDS)
Trimethoxyoctylsilane (TMOS)
Hexamethyldisilazane (HMDS)
Polydimethylsiloxane (silicone oil)

Hexadecylsilane (D16)
Octamethylcyclotetrasiloxane (D4)
Triethoxypropylaminosilane (TEPAS) + HMDS
Methyacrylsilane (MAS)

THE STATE UNIVERSITY OF NEW JERSEY
RUTGERS

Ceramic and Composite Materials Center

Surface Treatments & Attributes:

	Attribute to coating
- Dimethyl-dichloro-silane (DDS)	high OH conversion
- Trimethoxy-octyl-silane (TMOS)	long chain - high polar stability
- Hexamethyl-di-silazane (HMDS)	pH neutral
- Poly-dimethyl-siloxane (silicone oil)	highest hydrophobicity
- Hexa-decyl-silane (D16)	hydrophobic, wets into water
- Octamethyl-cyclo-tetra-siloxane (D4)	low residual HCl
- Triethoxy-propyl-amino-silane (TEPAS) + HMDS	highest pH silica - 9
- Methyacryl-silane (MAS)	functional

THE STATE UNIVERSITY OF NEW JERSEY
RUTGERS

Ceramic and Composite Materials Center

Lotus Effect® & Self - Cleaning Surfaces

Nelumbo nucifera
(Lotus Pflanze)

RUTGERS

THE STATE UNIVERSITY OF NEW JERSEY

Ceramic and Composite Materials Center

Comparison: Nature's model and artificial surface

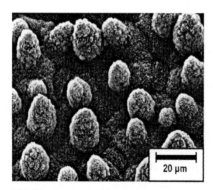

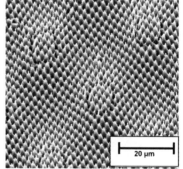

SEM images of *Nelumbo nucifera* and our first self-cleaning surface

RUTGERS

THE STATE UNIVERSITY OF NEW JERSEY

Ceramic and Composite Materials Center

Surface Effects: Lotus-Effect®

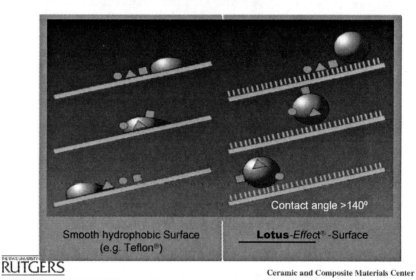

Non self-cleaning and self-cleaning surfaces

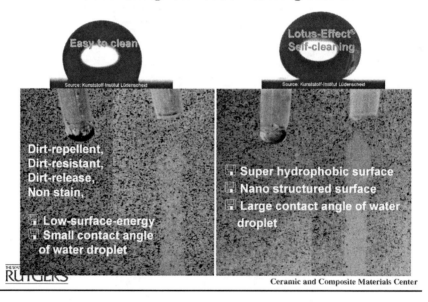

Summarizing Matching Up the "Tools" with the Tasks:

•A. Rheology
- SiO_2 7 – 12 nm core, treated / untreated dependent on details

•B. Free flow & Electrostatics
- SiO_2 12 – 40 nm core, Al_2O_3, treated / untreated dependent on details

•C. Scratch Resistance
- SiO_2 12 – 16 nm core, treated, structure modified

•D. Anti-corrosion, water repellency
- SiO_2, treated

•E. Lotus Effect® & self - cleaning surfaces
- SiO_2, treated

•F. Transparent UV Absorption & Conductivity
- UV – ZnO, Conductivity - In_2O_3 / SnO_2

RUTGERS

Ceramic and Composite Materials Center

Medical Applications of NanoComposite Particles as Drug Delivery Systems

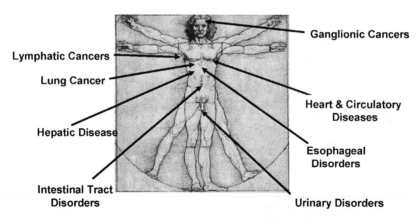

From Leonardo da Vinci (1492-1519): Man. Venice, Galleria dell'Accademia.

RUTGERS

Ceramic and Composite Materials Center

Nanomedical Applications
Example of Targeted Drug Delivery
Systems Based on NanoComposite Particles

Surface Functionalized NanoComposite Particles

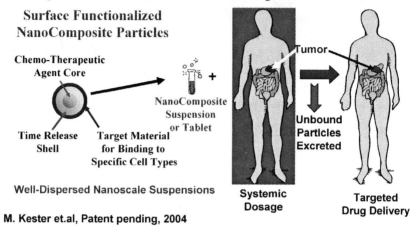

Chemo-Therapeutic Agent Core

NanoComposite Suspension or Tablet

Time Release Shell

Target Material for Binding to Specific Cell Types

Well-Dispersed Nanoscale Suspensions

Tumor

Unbound Particles Excreted

Systemic Dosage

Targeted Drug Delivery

M. Kester et.al, Patent pending, 2004

RUTGERS
THE STATE UNIVERSITY OF NEW JERSEY

Ceramic and Composite Materials Center

Hunter - Killers

Application - Bacterial, viral, and foreign body binding with cell death

Nanocomposite particles with cell membrane lysing agent core; resorbable shell or porous shell with resorbable coating; target functionalization for tumor, bacterial, viral or other foreign bodies

Example:

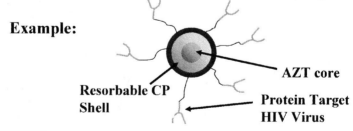

Resorbable CP Shell

AZT core

Protein Target HIV Virus

RUTGERS
THE STATE UNIVERSITY OF NEW JERSEY

Ceramic and Composite Materials Center

Example of Reverse Micelle Synthesis of Nanocomposite Particles with Silica Shells

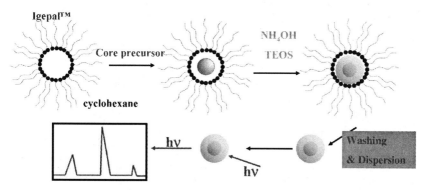

Ceramic and Composite Materials Center

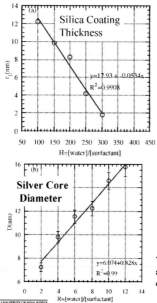

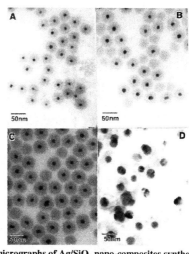

TEM micrographs of Ag/SiO$_2$ nano-composites synthesized at the conditions of X=1 and H=100, and different water content: (a) R=2; (b) R=4; (c) R=6; (d) R=10.

From Li, et.al, Langmuir, 1999
Ceramic and Composite Materials Center

Summary

•Nano-materials: ~ 10^{-9} meters (nano-meters, nm, 1/1,000,000,000 of a meter, 1 nm is ~ 3-4 atoms wide)

•Materials, structures or systems where behavior and control of processes are at the atomic or molecular level

•Materials that exhibit novel (unexpected) properties, behavior and phenomena

•Impact on the health, wealth, and security of the world's people is expected to be at least as significant as the combined influences in this century of antibiotics, the integrated circuit, and human made polymers

•Benefits to Society will be in areas of:

•Energy
•Environment
•Telecommunications
•Medicine
•Biology

•Computing
•Information
•Homeland Security
•Transportation

THE STATE UNIVERSITY OF NEW JERSEY
RUTGERS

Ceramic and Composite Materials Center

Ackowledgement of Support

The Ceramic and Composite Materials Center
The Department of Materials Science and Engineering, Stephen Danforth, Chair
The Center for Nanomaterials, Bernard Kear, Director
The Particulate Materials Center at Penn State, James Adair, Director
The National Science Foundation
The Army Research Laboratory
The Army Research Office
Degussa Corporation
Millennium Chemical Corporation
St. Gobain Corporation
Ceradyne Inc
Simula/Armor Holdings
Surmet
Others......

THE STATE UNIVERSITY OF NEW JERSEY
RUTGERS

Ceramic and Composite Materials Center

Energy and Raw Materials

THE ELECTRICITY MARKET & CRITICAL ISSUES

Lucia Harvey
Tennessee Valley Authority

Abstract
Many factors affect the supply of electricity for manufacturing; including growing consumer demand, limited new power plant construction and environmental impact issues. Record warm summers have been recorded in the last decade. Significantly strong hurricanes have damaged infrastructure in the Gulf region of the US. Environmental restrictions have increased in recent years regarding by-products from energy production. Weather conditions in other geographic areas affect the supply of natural gas (LNG) and increased demand in China and India and unrest in Africa and the Middle-East add to energy supply problems.

The outlook for energy generation for the next decade and beyond is highlighted. Demand and capacity are expected to continue their upward trends and reach equilibrium in about 2015; however the "reserve margin" may drop to 10% from a level of about 30% today.

Weather: The Past and the Present

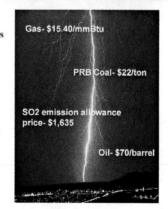

TVA 2005- Commodity Prices " Perfect Storm"

- 2005 warmest year on record
- Hurricanes Katrina and Rita damaged Gulf coast oil and gas infrastructure
- Joint line maintenance issues hampered PRB deliveries
- Sulfur dioxide emission allowances reached record highs
- Drought in Spain increased demand for Liquefied Natural Gas (LNG)
- Increased fuel demand in China and India increased global competition
- Political unrest in Nigeria and Iran threaten oil supply

Gas- $15.40/mmBtu

PRB Coal- $22/ton

SO2 emission allowance price- $1,635

Oil- $70/barrel

Business Sensitive   3

35

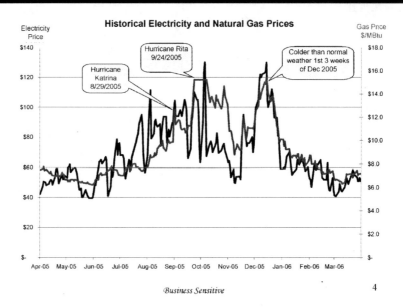

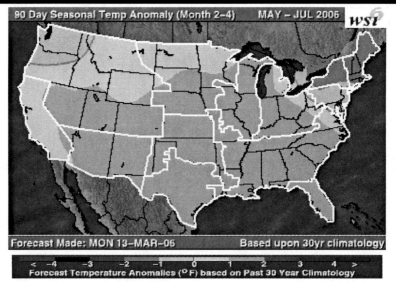

Precipitation Outlook

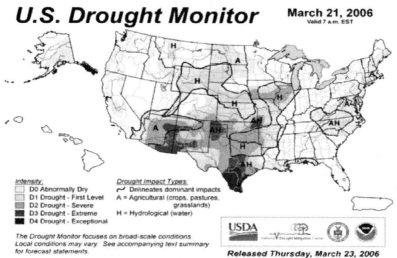

 Summer Temperatures and Loads

- **Warmer than normal temperatures**
- **Drought conditions continue across much of the central US**

These weather conditions cause
electricity prices to rise

Business Sensitive 7

TVA

Fuel and Electric Price Trends

TVA **Coal Trends**

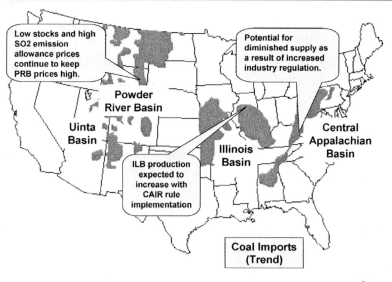

Low stocks and high SO2 emission allowance prices continue to keep PRB prices high.

Potential for diminished supply as a result of increased industry regulation.

Powder River Basin

Uinta Basin

Illinois Basin

Central Appalachian Basin

ILB production expected to increase with CAIR rule implementation

Coal Imports (Trend)

TVA

Coal Supply and Demand-TVA Region

- <u>CAP</u> prices high due to SO2 regulations and depleting coal reserves
- <u>ILB</u> production and demand expected to remain in equilibrium due to coal-on-coal competition
- <u>PRB</u> production steadily increasing to meet growing demand as transportation constraints alleviate
- <u>PRB</u> is expected to supply the majority of coal consumption growth
- <u>Uinta</u> production is expected to remain flat

Business Sensitive   10

TVA

Electricity Price Forecasts with Uncertainty

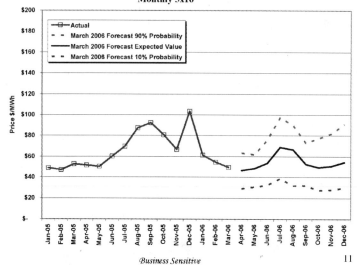

INTO TVA Actual and Forecasted Electricity Prices
Monthly 5x16

Business Sensitive   11

ⅣⅤ Electricity Price Forecast – Real $ Weekday

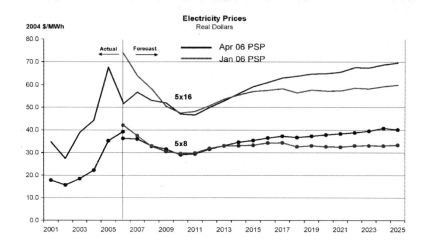

ⅣⅤ Forecasts For The Next 12 Months

- **Fuel and electricity prices are not forecasted to trend either up or down, but will show normal volatility**
  - On peak prices are set by combined cycle plants fueled by gas

- **Price movement of commodities (coal, allowances, gas, market electricity) have moderated, but at higher levels**
  - Forecast price declines are not significant…and may not occur

- **Fuel supplies (coal, gas) are recovering from recent problems, but at risk if new issues develop**

Supply and Demand

TVA Region at a Glance – Supply and Demand

- **546,558 MW of installed capacity**
  - Coal – 229,878 MW
  - Nuclear – 65,150 MW
  - Gas – 202,862 MW
  - Hydro – 29,392 MW
  - Fuel Oil – 11,769 MW
  - Other – 7,508 MW
- **Peak load forecast - 473,685 MW**
- **Reserve Margin - 20%**
- **10,992 MW under construction (NewGen)**
  - Nuclear - 1,280 MW
  - Coal - 3,069 MW
  - CC - 5,569 MW
  - CTs - 345 MW
  - Other – 728 MW

Business Sensitive 15

TVA Supply and Demand Balance Assumptions

- Demand outlook for the 5 region remains relatively unchanged with a 1.7% growth rate

- Supply and demand equilibrium remains at 2015 for determining entrance of new supply with a long-term reserve margin target of 10%

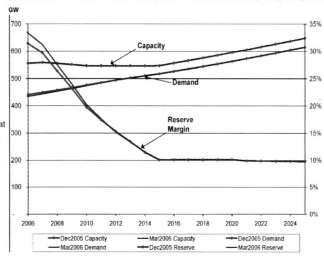

August 2005: Four Adverse Trends

August 2005 Supply Curve

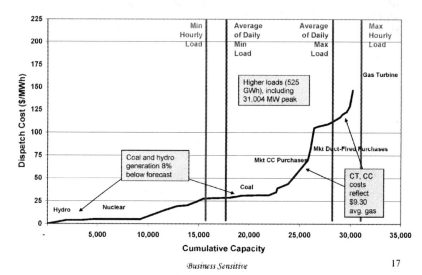

Business Sensitive   17

Capacity Decisions

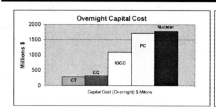

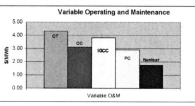

- Capacity decisions depend on the costs and operating efficiencies of different options, fuel prices, and the availability of Federal tax credits for investments in some technologies
- Natural gas plants are the least expensive to build but have the highest operating costs
- Coal and Nuclear plants are more expensive to build with lower operating costs
- The largest amounts of new capacity are expected in the Southeast and the West
- Coal-fired plants will make up the majority of the capacity additions through 2030 due to rising gas prices

EIA/Annual Energy Outlook 2006

Business Sensitive   18

TVA Power Supply Availability

- **Lower Hydro in forecast due to dry conditions and USACE operational changes on the Cumberland River**
  - Calendar year hydro generation is 65% of normal
- **Added purchases from several sources to address capacity needs**
- **Improvement in coal inventory situation**
- **No major outages planned for fossil or nuclear units**
- **CT generation available for economic dispatch**
  - Fast load start
  - Remote operations on Gas
- **Brown's Ferry 1 will come on line May 2007**

Business Sensitive   19

TVA Summary

- **Electricity prices about the same as last summer (pre-hurricanes!)**
  - Forecasted average on-peak price $65-70/MWh
  - 14% of Gulf supply remains shut-in with limited impacts to TVA due to mild winter
- **Transmission continues to be tight causing TLR's in the Eastern Interconnect**
- **Potential liabilities:**
  - Lower than already low hydro forecast
  - Fuel supply (coal or gas)
  - Long term forced outage of large thermal units

Business Sensitive   20

THE CURRENT LITHIUM SITUATION

Jason Woulfin
SQM North America

Abstract
World demand for lithium continues to grow, especially in several markets that are unrelated to porcelain enamel. The outlook for near term supply is discussed. Lithium is derived from mineral (spodumene, 1.5 to 4.0% Li) and lithium brines. Lithium brines have the highest concentrations of the element than any other natural source. Lithium carbonate is the largest product produced regarding percentage of production. Asian markets are growing and using lithium products at an ever increasing rate. Chile (Solar de Atacama) is the leading supplier of lithium chemicals. The geography of Solar de Atacama is superior to all other sites worldwide regarding extremely low rainfall and high evaporation rate. This combination results in a very energy efficient operation. It is anticipated that new production capacity, approximately 30% increase, will come on stream between 2005 and 2008.

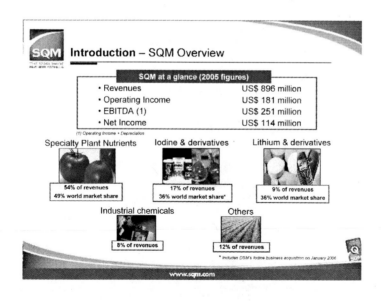

Introduction – SQM Overview

Caliche Ore

Nitrate

Iodine

Sulfate

Salar Brines

Potassium

Lithium

Boron

- Largest world reserves.
- Best grades.
- Good accessibility.

www.sqm.com

Introduction – Definitions

The world lithium industry can be divided in 4 segments, according to its added value level and characteristics for the products:

Lithium minerals
- Spodumene and other lithium minerals containing 1.5-4.0% Li, used directly in the glass and ceramic industry

Lithium Carbonate Merchant market
- Total demand for Lithium Carbonate, excluding its use in the captive production of LiOH, LiCl and Performance Chemicals

Lithium Basic Chemicals
- Lithium Hydroxide and Lithium Chloride

Lithium Performance Chemicals
- Butyllithium, Specialty Organics, Specialty Inorganics, Lithium Metal.

www.sqm.com

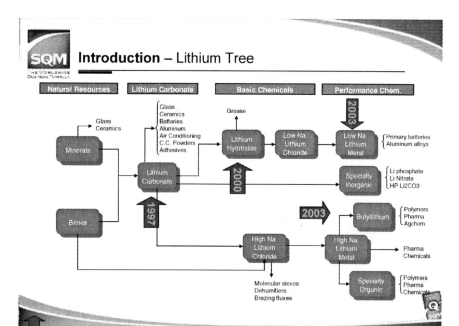

Demand – Apparent Consumption Last 12 Months

- **Demand for Lithium chemicals in 2005 reached 76.000 MT LCE, 6% higher than 2004.**

  - A healthy growth rate is expected for 2006 (+5%).

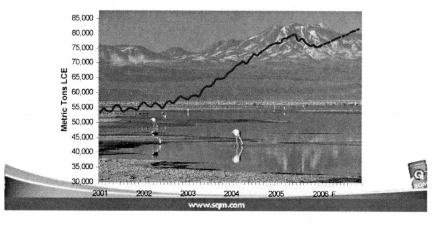

Demand

- **The largest portion of the demand corresponds to the merchant market for Lithium Carbonate (Li$_2$CO$_3$).**

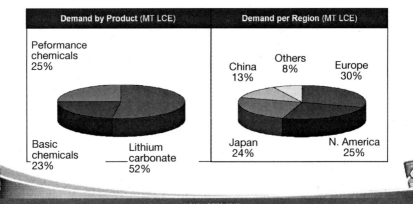

Demand – China

- **In the last years, Chinese demand for lithium chemicals have soared.**
- **Chinese demand will remain strong in the next coming years.**

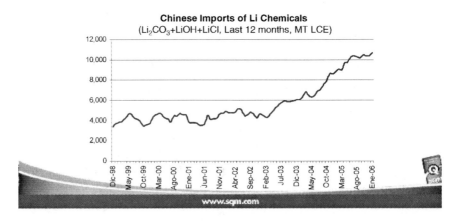

Demand – Forecast 2006 - 2010

- **Global demand for Lithium chemicals will increase at 5.000 MT/y or 5% per annum.**

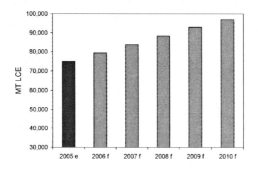

... **How this increasing demand will be satisfied?**

www.sqm.com

Supply – Sourcing of lithium

- **The entrance of SQM led a significant change in terms of the source of lithium.**

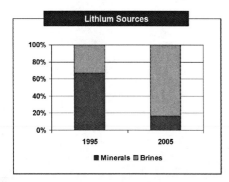

- In 1995, only Silver Peak (US) and SCL (Chile) produced lithium compounds using brines

- During the period 1997-1999 several Li mineral production sites either closed down or reduced their production: China, USA, Russia, Australia.

- In 2005, more than 85% of lithium production comes from brines.

www.sqm.com

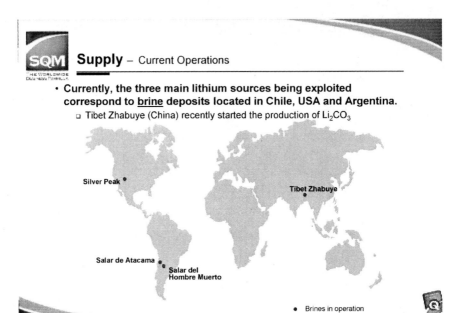

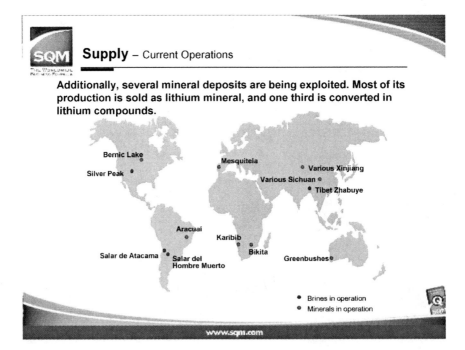

Supply – Market Share

- **SQM is the largest producer of Lithium Chemicals, with 36% market share.**
- **As a country, Chile has consolidated its leading position.**

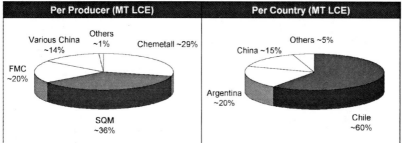

Per Producer (MT LCE)	Per Country (MT LCE)
Various China ~14%, Others ~1%, Chemetall ~29%, FMC ~20%, SQM ~36%	Others ~5%, China ~15%, Argentina ~20%, Chile ~60%

www.sqm.com

Supply – Ongoing Expansion Projects

New capacity is coming on stream.

- Chile will add 17.000 MT in the next two years.

  □ In a first stage SQM will increase its production capacity from 28.000 MT/yr to 40.000 MT/yr. In a second stage SQM's capacity will increase further to 60.000 MT/yr.

  □ SCL-Chemetall capacity is increasing by 5.000 MT/yr this year.

- **China has several brine and mineral projects that will materialize accordingly with market conditions.**

  □ Although they are currently producing small quantities, several projects have been announced: Tibet Region, Quinghai Basin and various mineral projects in Sichuan and Jiangxi provinces.

www.sqm.com

Supply – SQM Investment Program

SQM investment plan will reinforce our leadership in the lithium industry.

- SQM is undergoing a US$ 650 million investment program in 2005-2008 to increase its nitrates, iodine and lithium production by 30%

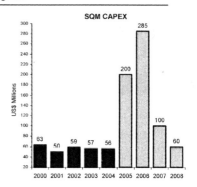

SQM CAPEX

www.sqm.com

 Supply – Salar de Atacama

○ Cyprus Foote Minerals started production of Li_2CO_3 in the Salar de Atacama in 1984.

○ SQM started its Li2CO3 production in 1997, as a by product of the production of KCl.

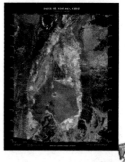

www.sqm.com

Reserves – Salar de Atacama

•The Salar de Atacama has remarkable advantages compared to other salt lakes

Brine Composition

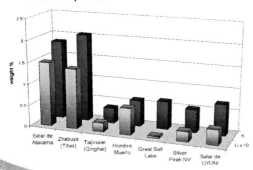

Rainfall
mm/year

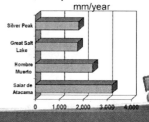

Reserves – Salar de Atacama

- SQM's mining rights: 1.950 km^2, that encompass 85% of the mining rights of the Salar de Atacama.
- SQM's proven reserves in the upper 40 m of 50% of its mining rights amount to over <u>11 million tons of lithium carbonate</u>.

Summary

- Since the entrance of SQM, significant changes have occurred in the Lithium market, in terms of supply sources.
- Currently, the Salar de Atacama is the world largest source of lithium.
- The Salar's leadership position should continue in the medium and long term, because of its remarkable characteristics:
  - ❑ Quality and abundance of the reserves
  - ❑ Advantageous natural operational conditions
  - ❑ Good accessibility and simple logistic
  - ❑ Combined production with other chemicals
- Further expansions in SQM's production capacity can be accomplished in the future to meet demand growth in current and new applications for lithium.
- The Salar de Atacama ensures the industry that lithium will be abundantly available for many years to come.

www.sqm.com

ENERGY SAVING RECOMMENDATIONS – PANEL DISCUSSION

David Latimer
Whirlpool Corporation

Peter Vodak
Engineered Storage Systems

Mike Horton
KMI Systems, Inc.

Abstract
Energy savings is critical to continuing profitable operations of enameling plants as the cost of energy of all types continues to increase and in some instances escalate to higher and higher levels. Washers, dryers, environmental rooms (air conditioned), furnaces, cooling tunnels and make-up air all require energy. Heat energy is the largest component of the energy costs for enameling plants. Reduction of processing temperatures of about 20°F is shown to decrease heating costs by nearly 50% and greater savings with decreased of about 40°F for the parts washers. Elimination of air infiltration by air seals also dramatically reduces energy costs of dryers, dry-off ovens and furnaces. A patented air vestibule design is illustrated as well as furnace air seals which reduce air exchange at process entrances. Reduction of temperatures nearly 400°F at the entry point of furnaces is illustrated. Excess furnace heat is redirected to the enamel dryer and cooling tunnel heat is returned to various plant locations with associated costs savings.

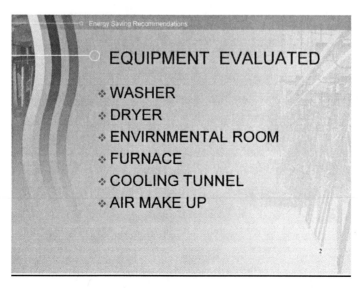

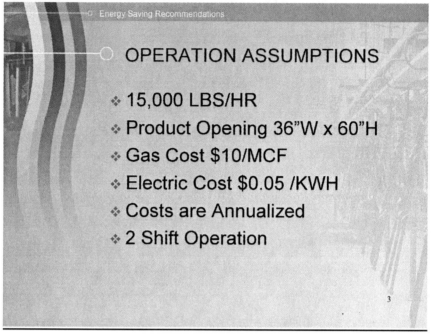

Energy Saving Recommendations

OPERATION ASSUMPTIONS

- ❖ 15,000 LBS/HR
- ❖ Product Opening 36"W x 60"H
- ❖ Gas Cost $10/MCF
- ❖ Electric Cost $0.05 /KWH
- ❖ Costs are Annualized
- ❖ 2 Shift Operation

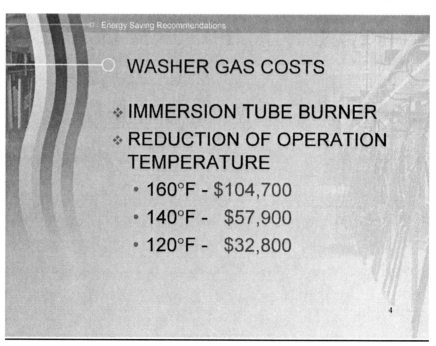

Energy Saving Recommendations

WASHER GAS COSTS

- ❖ IMMERSION TUBE BURNER
- ❖ REDUCTION OF OPERATION TEMPERATURE
  - 160°F - $104,700
  - 140°F - $57,900
  - 120°F - $32,800

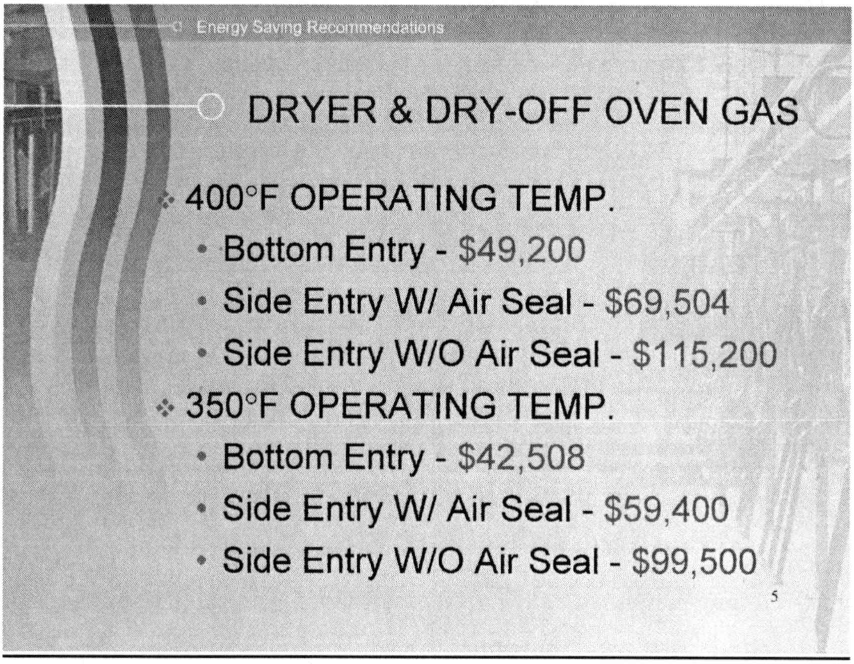

Energy Saving Recommendations

DRYER & DRY-OFF OVEN GAS

400°F OPERATING TEMP.
- Bottom Entry - $49,200
- Side Entry W/ Air Seal - $69,504
- Side Entry W/O Air Seal - $115,200

350°F OPERATING TEMP.
- Bottom Entry - $42,508
- Side Entry W/ Air Seal - $59,400
- Side Entry W/O Air Seal - $99,500

5

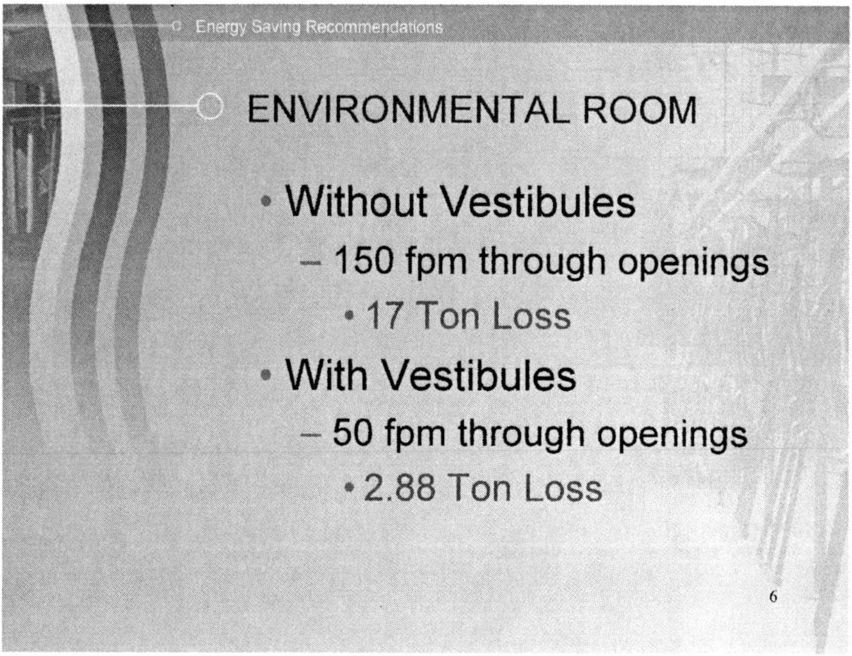

Energy Saving Recommendations

ENVIRONMENTAL ROOM

- **Without Vestibules**
  - 150 fpm through openings
    - 17 Ton Loss
- **With Vestibules**
  - 50 fpm through openings
    - 2.88 Ton Loss

6

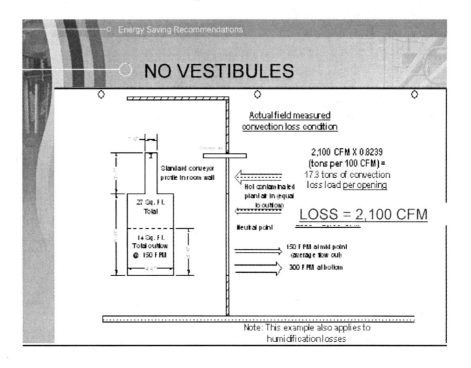

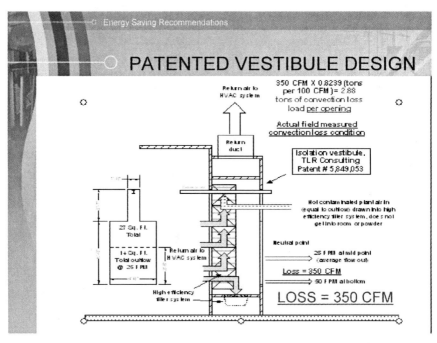

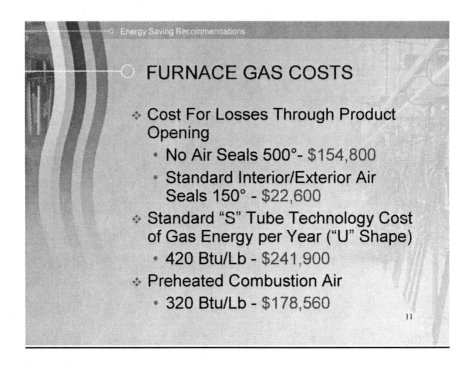

FURNACE ENTRANCE

SURFACE TEMPS WITH NEW KMI AIR SEAL SYSTEM.
OLD AIR SEAL TEMPS SHOWN IN PARENTHESIS

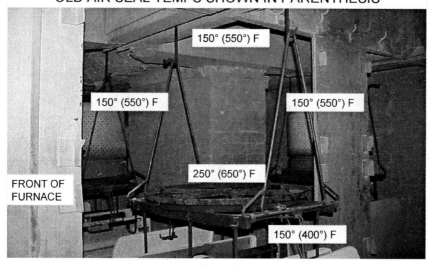

FURNACE ENTRANCE

NEW AIR SEAL SYSTEM ENTRANCE AIR FLOWS

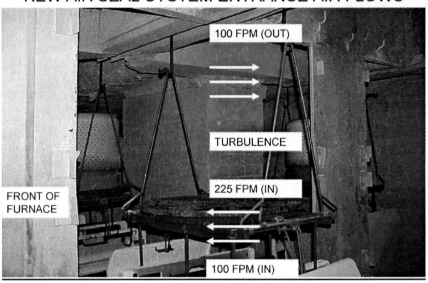

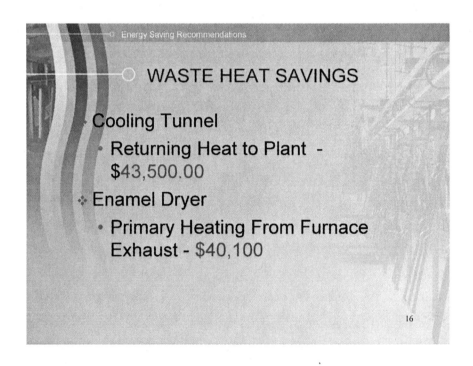

Laboratory Investigations

GLASSY SURFACE FUNCTIONAL MODIFICATION BY NANO-MODIFIED SOL-GEL TECHNOLOGY*

Christian Schlegel
Pemco Corporation

Abstract
An overview of new sol-gel technology: Nano-Clean was presented along with the property enhancement this new technology brings to glass surfaces, specifically porcelain enamel coatings.

Various methods have been used to apply thin property enhancing coatings on glassy substrates: chemical vapor deposition (CVD), physical vapor deposition (PVD) and sol-gel processing. Sol-gel has a number of added benefits compared to other processes, including low energy input necessary for curing. Processing of sol-gel may be accomplished by spraying, dipping, flow coating, spin coating, roll coating or silk screening. The coatings produced have to be handled carefully to avoid cracks and other defects during the drying process. The resultant surfaces may be hydrophobic or hydrophilic and have specific properties. Anti-graffiti, anti-fogging and colored coatings are possible. Self-cleaning effects have also been demonstrated on glass surfaces such as windows, allowing rain water to wash away soils. On porcelain enamels, baked on soils are more easily removed with the Nano-Clean surface treatment compared to untreated pyrolytic enamel surfaces.

PEMCO

➢ Glassy surfaces:

- Glass
- Ceramics
- Porcelain enamels

don't always have the desired surface properties.

➢ Therefore, it is sometimes desirable to apply on glassy surfaces an additional functional coating, which can utilize the following technologies:

- CVD (Chemical Vapor Deposition)
- PVD (Physical Vapor Deposition)
- SOL-GEL Processing

SCHC-2

*Presentation given by Pat Pawlicki.

PEMCO

➢ It is Sol-Gel technology and the numerous additional benefits that I would like to
share with you, especially:

- What is meant by a "Sol-Gel"-Coating.

- Some additional & achievable surface properties through Sol-Gel coating.

- And more specifically in connection with enamelled surfaces:
  "NanoClean", a hybrid "sol-gel" coating, yielding anti-stick properties.

SCHC-3

PEMCO

➢ The Sol-Gel technology enables the transformation of molecular precursors, reacting
with water, together with oxide materials, a "Sol"-Phase (colloidal) and a "Gel"-
Phase (solid material, whose pores contain a liquid phase).

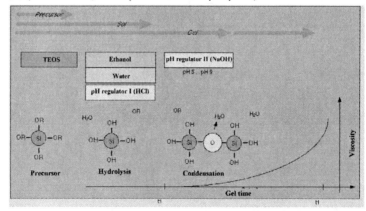

SCHC-4

PEMCO

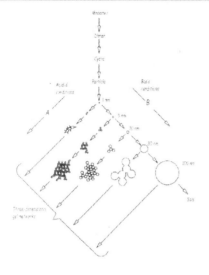

> The reactions, that are leading to gels are complex and are depending on many parameters:

- type of molecular precursor
- proportion: water : precursor
- type of catalyst   (alkali/acid)
- pH-value
- type of solvent used
- additives (activators, electrolyte, tensio-active)
- temperature
- concentration of the precursor

SCHC-5

PEMCO

> The gels that are produced this way, can be applied as follows:

| Spraying | Dip-Coating | Flow-Coating | Spin-Coating | Roller-Coating | Silk-Screening |

PEMCO

> The bonding mechanism of Sol-Gel layers on glassy surfaces is based on the presence of "OH"-Groups on the surface.

> To avoid decomposition or crack formation, the gel needs to be very carefully dried.

> Possible layer thickness:

  5-8 Nanometers

> Higher thicknesses are possible through the repetition of several successive coatings and drying steps.

> A densification of the gels can be achieved by an annealing step (min. 400°C).

SCHC-7

PEMCO

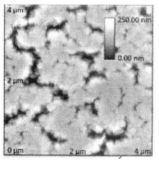

> The porous structure of gels allows for nano-sized particles to be located in the pores. This material can be chemically bound or not to the Sol-Gel matrix.

> This property is opening many new ways and particularly the way to multi-functional materials.

> This nano-sized material can be organic or inorganic. In the case of an organic material, the Sol-Gel layer becomes a hybrid material (ORMOCER = organically modified ceramics).

> The fact that the layers of Sol-Gel materials are generally nano-sized, most of them are transparent and optically invisible.

PEMCO

➢ A SiO$_2$ Sol-Gel coating is notably improving the chemical resistance of a glassy surface e.g. enamel, ceramic coatings.

➢ A SiO$_2$ hybrid Sol-Gel coating on a glassy surface (ceramic, enamel, etc.

↑ Makes the "graffiti" elimination easier.

↑ Discourages the "tagger" from creating "graffiti" (a permanent result is no longer possible).

Non-Treated

Treated

SCHC-9

PEMCO

➢ A hydrophilic TiO$_2$ Sol-Gel coating enables an interesting anti-fogging effect on glassy surfaces (mirrors, automotive glass).

➢ A SiO$_2$, nano-silver modified Sol-Gel coating proposes real anti-bacterial properties.

Nanosilber tötet Bakterien

SCHC-10

PEMCO

> A Sol-Gel Coating modified nano-film layer using mineral pigments enables the development of transparent colors (majolica effect).

> A TiO_2 -based Sol-Gel coating is leading to photo-catalytic (light sensitive) glassy surfaces with self-cleaning properties.

SCHC-11

PEMCO

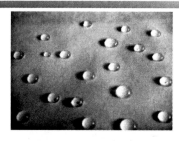 

> The possibility of giving anti-stick properties to enamelled or glaze surfaces by coating them with:

NanoClean

a hybrid SiO_2-modified Sol-Gel layer.

SCHC-12

PEMCO

| "NanoClean" |

1. "NanoClean"/Structure :

 Step 1

Sol-Gel Solution

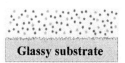

Glassy substrate

- Component, which chemically reacts with the glassy surface.
- Inorganic or organic network improving abrasion resistance.
- Component, which makes the surface water- & oil-repellent.

 Step 2

Elimination of the solvent through evaporation

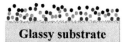

Glassy substrate

↑ First molecular organisation

 Step 3

Glassy substrate

↑ Final molecular organisation with the creation of an anti-stick layer.

SCHC-13

PEMCO

| "NanoClean" |

2. "NanoClean"/Action : "Easy-To-Clean" surfaces need to be at the same time hydrophobic and oliophobic

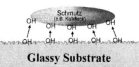

Glassy Substrate

- hydrophilic surface

- high surface energy

- smooth surface

↑ **Difficult to clean**

Glassy Substrate

- hydrophobic/oliophobic surface

- low surface energy

- rough surface

↑ **Easy to clean**

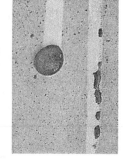

SCHC-14

PEMCO

> "NanoClean"

3. "NanoClean"/Surface roughness :

> Defined through AFM-Measurements.

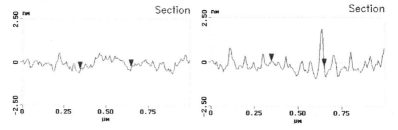

Non-treated surface
Low R-value
↑ numerous contact points

With NanoClean treated surface
High R-value
↑ few contact points

SCHC-15

PEMCO

> "NanoClean"

3. "NanoClean"/Surface roughness :

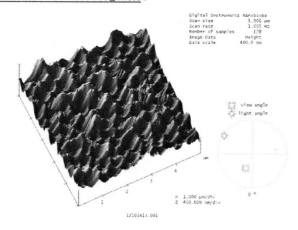

SCHC-16

PEMCO

> **"NanoClean"**

4. "NanoClean"/Hydrophobic/Oliophobic-definition :

> Through the contact angle

- of a water drop (α) (Definition of hydrophobic)
- of a hexadecane drop (α_1) (Definition of oliophobic)

Glassy surface

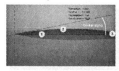

$\alpha \leq 40°$ ↑ hydrophilic   Contact angle $\alpha \geq 105°$↑ hydrophobic

$\alpha_1 \leq 20°$ ↑ oliophilic $\alpha_1 \geq 65°$↑ oliophobic

SCHC-17

PEMCO

> **"NanoClean"**

5. "NanoClean"/Properties :

> **High scratch and abrasion resistance (Hydrophobic-loss of 1 to 3% after 10,000 cycles).**
> **Optically neutral**
> **UV-resistant**
> **Water- and Oil-repellent**
> Solvent-resistant
> Improves acid resistance
> Does not influence alkali-resistance
> **Heat resistance: up to 300°C (570°F)**
> Food-neutral (Food Approval Certificate...)
> Environmentally friendly

SCHC-18

PEMCO

"NanoClean"

6. "NanoClean"/Coating process :

Pre-cleaning	→	Coating	→	Curing	→	Post-cleaning

➤ **Pre-cleaning**
- The glassy surface must be extremely clean (grease- and dust-free).
➤ **Coating**
- Different possibilities: Wiping - Dipping - Spraying (Aerosol)
➤ **Curing**
- "Initial Curing": 10 Minutes at Room temperature or 2 Minutes at 80°C (176°F)
- "End-Curing" : after 24 hours (Handling possible after "Initial Curing").
➤ **Post-cleaning**
- Depends on the type of coating.
- Elimination of non-reacted monomers or spraying spots.

SCHC-19

PEMCO

"NanoClean"

7. "NanoClean"/AHAM Test/Easy-To-Clean Test Method :

➤ **Mixing of 9 different foods:**

- Ground beef; grated "Cheddar"-cheese; homogenized whole milk; powder sugar; cherry juice; instant tapioca; 1 raw egg; flour; tomato juice.

➤ **Testing procedure:**

- The test plate surface is soiled with the mixture and installed in the middle of the baking oven
- Baking cycle 20 to 205°C (68 to 401°F) + dwell time = 1 hour.
- Natural cooling.

PEMCO

"NanoClean"

8. "NanoClean"/AHAM Test/Easy-To-Clean Test Method :

> ### Dirt removal through:

- a plastic scraper
- a plastic scouring pad
- a copper scouring pad

> ### Easy-To-Clean evaluation:

0	=	no residues	0 %
1	=	a small amount of residues	10 %
2	=	a few residues	25 %
3	=	average residues	50 %
4	=	a lot of residues	75 %
5	=	all residues are remaining	100 %

SCHC-21

PEMCO

"NanoClean"

8. "NanoClean"/AHAM Test/Easy-To-Clean Test :

Cleaning means	A Plastic scraper						B Plastic scouring pad						C Copper scouring pad					
Number of AHAM test cycles	1	2	3	4	5	45	1	2	3	4	5	45	1	2	3	4	5	45
1. Black standard	4	4	5	5	5	/	4	4	4	4	4	/	1	1	1	1	2	/
2. Black treated	0	0	0	0	0	0	0	0	0	0	0	0	0	0	0	0	0	0
3. Grey standard	3	3	5	5	5	/	3	3	4	4	5	/	0	0	1	1	2	/
4. Grey treated	0	0	0	0	0	0	0	0	0	0	0	0	0	0	0	0	0	0
5. Teflon	0	0	0	0	0	0	0	0	0	0	0	0	0	0	0	0	0	0

Rest of soils

0	= 0 %	Easy-to-clean
1	= 10 %	
2	= 25 %	
3	= 50 %	
4	= 75 %	
5	= 100 %	Difficult-to-clean

Conclusion:
NanoClean treatment
= teflon
⇒ very good anti-stick properties

SCHC-22

PEMCO

> ## "NanoClean"

9. <u>"NanoClean"/Technical Data</u> :

- **"NanoClean" is a 2-component-system**

    * Activator: shelf life: 1 year
    * Base-solution: shelf life: 2 years
    * "NanoClean"-solution: shelf life: 2 weeks
    * Application: 10 to 15 ml/m^2

- **"NanoClean"/Cost impact**

    * $2.75 – $4.00/m^2 ($0.25 - $0.40 / ft^2

SCHC-23

PEMCO

> ## Conclusion

➤ <u>Positives</u>:

    - the numerous possibilities of surface property modification.
    - the simple way of application.
    - the low energy necessary for layer densification.

➤ <u>Negatives</u>:

    - the limited mechanical properties as a result of the nano-thickness of the layer (very little mass.
    - the highly accurate pre-cleaning needed (if the coating is not applied directly after the firing step).

➤ Nevertheless, the chemical Nano-technology remains a real break-through to improve or extend the properties of glassy surfaces.

SCHC-24

UPDATE ON THE STATUS OF PORCELAIN ENAMEL POWDER APPLICATIONS WORLDWIDE

William D. Faust
Ferro Corporation

Abstract

Electrostatic powder application continues to be a desirable process being utilized by appliance manufacturers in all parts of the world for many "traditionally" wet applications due to its economic benefits and improving flexibility. This paper presents an overview of electrostatic powder production technology in current use around the world.

Introduction

Electrostatic porcelain enamel powder application was experimented with in the early 1960's[1]. However, significant progress was not made until the early 1970's with the improvement of application equipment which was geared to organic powder coatings and the development of improved surface treatments of the glass powders. Organic powders have inherently high electrical resistivities, typically 10^{14} ohm-meters or higher. Glasses have resistivities of about 10^8 to 10^9 ohm-meters in bulk form. The proprietary glass treatments developed to coat the glass particles with a thin molecular layer of an organic type coating created resistivities in the 10^{14} ohm-meters range or higher. These resistivities allow the powders to retain a static charge for periods long enough to allow application, transport and subsequent firing of the powdered glass layer.

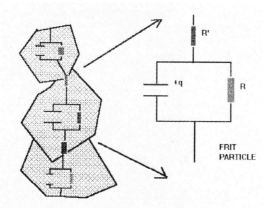

Figure 1 - Dielectric Material, Equivalent Circuit, "Industrial Electrostatics", Chap. 2, Fig. 2.34-b, p. 61. Research Studies Press, © 1994.

Adaptation of the techniques and equipment of organic powders required some engineering changes to accommodate the more abrasive and higher density glass powders. The impetus was the promise of lower costs by improving material utilization, less energy consumption and overall less labor.

Much of the early work on electrostatic powders was directed at understanding and controlling the various characteristics such as resistivity, fluidity, spray rate, color control, application control, eliminating or minimizing back emission [starring], particle size control, recirculation control, "trouble-shooting" in production operations, and many other things. As the technology progressed, the metal preparation requirements became less stringent with the development of no nickel-no pickle ground coats which resulted in the development of the two coat – one fire systems. Today we have a very efficient and clean system of enameling high quality products for consumer markets.

Powder Application Today
In the late 1970's and early 1980's, experimentation with a variety colors, almond, gold, ivory and brown was done[2]. Some of this work involved the use of oxides added to the powders. Long term use indicated that non-smelted colors were prone to segregation and did not continue in the market. However, ground coats with the traditional fleck appearance such as pyrolytic enamels were successfully developed for ranges as well as ground coats for rotisseries and water heater tanks.

Today we see electrostatic enamels mirroring the conventional wet systems for range and laundry applications, water tanks, barbecues, stove grates, and many other applications. Colors remain limited due to electrostatic application characteristics of particles of different sizes. Many of the pigment and other additions normally found in wet systems are not present in dry powder systems as their charge to mass ratios cause separation during the spraying process. The fine particles of pigment for example, 4 to 5 milli-microns in diameter, have much higher associated charges than the 25 to 25 milli-micron particles of glass. If uncontrolled, the finer particles will migrate to points of higher field strength, edges and embossed points and leave a non-uniform appearance. Similar effects will occur with added fillers in the glass powders.

Experience with electrostatic powder has allowed for increased flexibility in application such as dry-over-wet in some instances and the two coat-one fire process of demanding applications such as range tops, laundry spinner baskets and one-coat pyrolytic systems meeting the high heat and stress requirements of self-cleaning ovens.

Equipment Changes
The application equipment for electrostatic powders in the last 30 years has become more sophisticated and compact and in some cases simpler and easier to use. The first powder systems used cyclones and baghouses together. These required a significant amount of floor space. The gun movers were mechanically controlled and the guns were limited to voltages changes with less control over the current flow. Early experiments with powder equipment pointed out the abrasive nature of the glass powders. Hardened steel components were initially produced and eventually ceramic components that we use today as the steel parts were easily worn away.

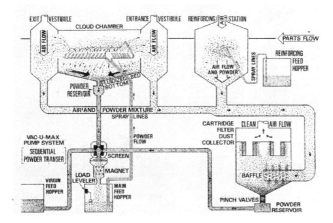

Figure 2 – DeVilbiss Cloud Chamber System, "Production Experience With Electrostatically Applied Porcelain Powder, R. G. Rion and E. Smithberger, Porcelain Enamel Institute Technical Forum, Vol. 38, 1976 p. 69-74.

Spray booths have changed significantly also. Sheet metal cabins originally designed had a tendency to accumulate significant amounts of fine powder on wall surfaces. This was due to the metal walls being grounded and attracting and holding the powder. This powder was typically very fine due to drift from the spray cloud. As the material was brushed back into the powder stream, the load of fines upset the particle size distribution of the powder and subsequently altered the application on the parts. Migration to "plastic" spray booth walls resulted in very low amounts of fine particles sticking to the walls and less dramatic swings in the particle size distribution of the batch being sprayed. Production stoppages to clean off the walls were essentially eliminated due to limiting the current drain through the wall surfaces, unlike that of metal surfaces.

Control of the operation on the guns for powder spraying has also undergone significant changes with the use of sophisticated computer interfaces. The voltages and currents are controlled with a greater degree of accuracy. Multiple guns may be controlled by integrated systems which can be preset for specific types of substrates being coated and can follow the parts for coating such as inverted oven cavities.

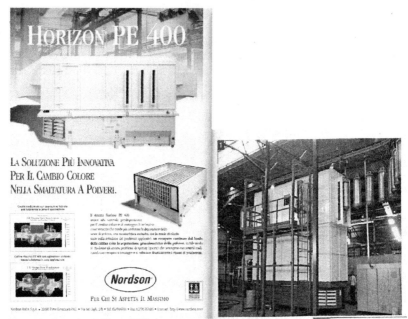

Figure 3 – Example of "Plastic Wall" Spray Booths - Nordson Corporation Spray Booth for Electrostatic Powder, Smalto Porcellanato, May-August 2005 and Boiler Coating Installation (South Africa), Smalto Porcellanato, September-December 2003, page 45.

Figure 4 – Examples of Integrated Spray Booth - Gema Electrostatic Powder Spray Booth, Smalto Porcellanato, September-December 2003.

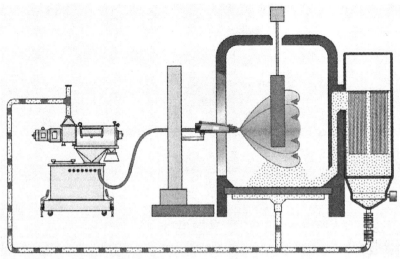

Figure 5 – Schematic of Powder Application System. Smalto Porcellanato. Figure 1- Smaltura elettrosttica a polveri:gli impianti, September-December 2000, page 159.

Use of Electrostatic Enameling

Electrostatic powder porcelain enamel application has now been widely adapted with plants in all major markets are using this process.

North America
       Canada – Ranges, Job shop applications
       Mexico – Ranges [two coat-one fire tops and pyrolytic oven interiors], grates
       United States – Ranges, laundry, barbeques, job shops [range components,
              grates of various types]

South America
       Argentina – Ranges
       Brazil – Ranges, Water Heaters
       Columbia – Ranges, water heaters and other appliances
       Uruguay - Water heaters

Asia-Pacific
       Australia
       Acid resisting ground coat for flatware (ranges)
       Two coat –one fire for range tops and doors
       Job shop enameling

New Zealand
Two coat-one fire and two coat-two fire processes for cooktops
Pyrolytic powder for oven cavities

Europe and Eastern Europe
60 to 100 Producers – Ranges, laundry, water heaters, grates
Acid resisting ground coats
Self-cleaning enamels
Heat exchanger coatings
White cover coat
Others miscellaneous applications

China
10 to 12 Producers
Water heaters
Barbecues
Architectural panels
Ranges

South Africa
Water heaters
Acid resisting black cover coat for appliances

Future
Electrostatic powder application of porcelain enamel should continue to see new uses and users worldwide. Markets that are developing in South America, Eastern Europe, Africa and Asia will most likely see the largest growth as these economic zones see rapid expansion. Development of improved electrostatic powders will continue to be important for producers and users alike.

References

[1] R. B. Reif, "Basic Principles of Electrostatic Coating", Porcelain Enamel Institute Technical Forum, Vol. 28, 1966, pages 113-120.

[2] W. D. Faust and D. R. Dickson, "Porcelain Enamel Powder Coatings: A Look at Colored, Pyrolytic and Water Heater Systems", Porcelain Enamel Institute Technical Forum, Vol. 38, 1976, pages 53-57.

THE USE OF SPECIALIZED ENAMEL COATINGS TO BOND CONCRETE TO STEEL

Philip G. Malone, Charles A. Weiss, Jr., Donna C. Day, Melvin C. Sykes, and Earl H. Baugher
U.S. Army Corps of Engineers
Engineer Research and Development Center
Vicksburg, MS

Abstract
Concrete and steel typically form weak bonds when steel reinforcement is used in concrete. A coupling layer fused to the surface of the steel should be capable of improving the ability of the steel to adhere to the surrounding concrete. Pull-out tests done on smooth steel rods that had been coated with porcelain enamel and also by enamel that was altered by the addition of mineral additives indicate that the mineral additives can increase the force require to extract the rods from a Portland cement mortar embedment. Small increases in maximum pull-out force occurred with enameled test rods abraded with Portland cement. An increase in maximum pullout force that amounted to three times that obtained with bare steel was observed when Portland cement was fused onto the enameled steel. Porcelain enamel is recognized as an excellent method of protecting steel from corrosion. Specially formulated porcelain enamel may serve the dual purpose of protecting the steel reinforcement from concrete and improving the bond of the steel to surrounding concrete.

Introduction
Steel is used for reinforcement in concrete structures in the form of steel fiber, rebar, welded wire mesh, and corrugated steel. Two problems arise in using steel reinforcement: the adhesion bond strength between concrete and steel is generally low and the concrete will crack when the steel starts to oxidize. Porcelain or vitreous enameling is widely recognized as providing excellent protection for steel. It should be possible to use the porcelain glass bonded to the steel as a coupling layer that can assist in improving the adhesion of concrete to steel.

Regardless of the type of reinforcement used, when steel is in contact with concrete the adhesion bond strength between the concrete and steel is low (Gray and Johnston, 1978), typically, on the order of 200 to 300 psi (1.4 to 2.0 MPa). While there is wide variability depending on the test apparatus and the conditions of preparation of the mortar there is generally little chemical adhesion between the hardened cement paste and clean, smooth, undeformed steel. Much of the problem of developing a strong chemical bond between concrete and steel surfaces relates to the formation of a thin layer of calcium hydroxide (portlandite) between the steel and the calcium silicate paste in the concrete (Al Khalaf and Page, 1979; Bentur, Diamond and Mindess, 1985). The goal of the current research is to investigate the ability of an enameled steel surface that contains a hydraulically reactive component such as Portland cement clinker to produce a strongly bonded interface between steel surfaces and the surrounding mortar (Fig. 1).

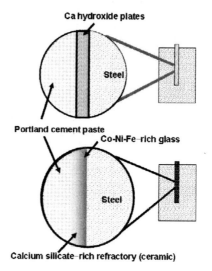

Fig. 1. Diagram showing the usual calcium hydroxide layer
at the interface and the approach of using a two-component
glassy enamel layer as an interface.

Ideally the glassy bonding layer would be a base-coat enamel that contains an outer layer with a mineral phase, for example a calcium silicate (crystalline and glassy phases) that would bond to the hydrated calcium silicate in the surrounding cement paste (Fig. 2). In this investigation, the effect of using an enamel coating along with various mineral phases to improve the bond strength was assessed. The study included an examination of plain enamel, (smooth and abraded) and enamel with either non-reactive or reactive phases.

Methods and Materials
The strength of the bond developed between treated steel and mortar was determined using a test procedure based on the pull-out test described in ASTM A 944-99. A standard low-carbon steel rod that was 76-mm (3 in.) long, and 6.35-mm (0.25 in.) in diameter was selected as the steel test surface. Bare steel control rods and six surface treatments were included in the testing program.

Control rods were prepared by cleaning the rods with a commercial metal cleaner and rinsing in alcohol and distilled water. The porcelain enameled rods were cleaned and enameled using Thompson (Newport, KY) 2600 series enamel frit fired at 760 to 815 °C (1400 to 1500 °F). Metal cleaning and preparation followed conventional methods (Society of Manufacturing Engineers, undated).

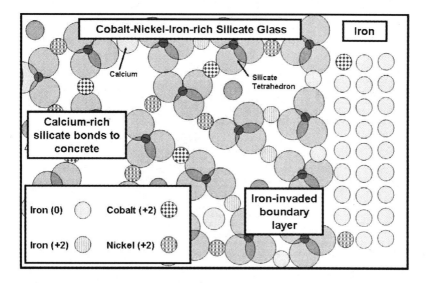

Fig. 2. Concept sketch showing a bonding layer that is an iron-rich silicate at the iron interface and a calcium-rich silicate at the concrete interface.

Three sets of enameled test rods with only the initial firing were set aside for testing. One set of the enameled-only rods was left with a smooth surface, a second set was abraded with fine grit polishing cloth, and a third set was abraded and dipped into dry Portland cement.

Three additional sets of test rods were prepared by firing additional mineral material into the enamel. After an initial firing, the three sets of rods were each coated with the selected additives (silica fume, muscovite, or Portland Type I-II cement) and reintroduced into the furnace to fuse the mineral material into the glass (Fig. 3). After firing and cooling the enamel-treated and the uncoated (control) rods were cast into mortar cylinders.

Casting was done with each test rod centered vertically rod in a 50.8 mm wide by 101.6 mm (2 inches wide by 4 inches long) cylinder mold. Each rod was set to depth of 66 (2.5 in.) mm. The mortar mixture used was the standard mortar mixture specified in ASTM C 109/C 109M-020. All samples were vibrated to consolidate the mortar during casting. The Portland cement used to prepare the mortar (Lone Star Type I-II, Brandon, MS) developed an unconfined compressive strength of 19.3 to 20.7 MPa (2800 to 3000 psi) when moist cured for 7 days. All samples were tested after 7 days. Unconfined compression strength tests were run on mortar cylinder to assure that the unconfined compressive strength of the mortar was within the standard range.

The adhesion bond strength was determined by measuring the force required to lift the test rod out of the cylinder using a pull-out test configuration used in Tattersall and Urbanowicz, 1974.

The pull tests were done using a Materials Testing System (MTS) Testing instrument (Eden Prairie, MN). All treatments were assessed using three identically prepared samples. The relative strengths of the bonds were compared by determining the average maximum vertical force required to move the rod.

Fig. 3. Test rod coated with enamel and fired a second time after Portland cement was added to the coating. The cement covers approximate one-half of the surface. Scale is in inches. One inch equals 25.4 mm.

Results

The results of the pull-out test of the control and single-fired test rods are presented in Fig. 4. The smooth enamel surface and the abraded enamel surface showed only minor differences in force required to lift the rod out of the cylinder. When unreacted Portland cement was applied to the rod prior of embedment the bond improved. All of the enameled test rods were removed with the enamel intact.

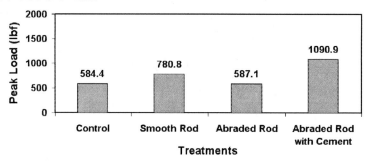

Fig. 4. Comparison of the peak load required to move the embedded rods. Each value is an average of three samples. The control rod is bare steel. The smooth rod is enamel coated but unabraded. Note: 1 pound force = 4.45 N.

The results of the pull-out test on the enameled rods that had been fired to fuse on a mineral additive are shown in Fig. 5. The Portland cement coated rod required over three times the force required to move a bare steel rod. The enamel with silica fume increased the adhesion slightly. Moving the he rod with fused mica in the enamel required over twice the force required for the bare steel (control) rod.

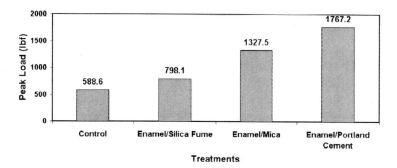

Fig 5. Comparison of the peak load require to move the embedded rods with the mineral additives. Note: 1 pound force = 4.45 N.

Fig. 6 shows a test cylinder make with a enameled rod that was made by fusing Portland cement into the enamel. The mortar cylinder was cracked after the test to allow the interface between the rod and mortar to be examined. The enamel was stripped off of the steel and remained bonded to the surface of the mortar.

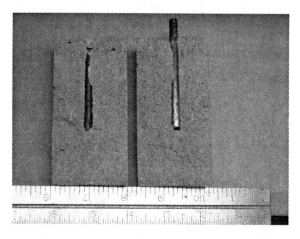

Fig. 6. Both halves of a test cylinder that shows that the enamel coating on the steel was stripped off during the pull-out test. Cracking of the mortar during pull-out was noted only at the very top of the embedded rod.

Conclusions

The study undertaken here indicates that:

a) Coating steel with conventional enamel does little or nothing to increase the adhesion bond between the enameled steel and surrounding portland cement mortar,

b) Small improvements can be obtained if the exterior glassy surface is abraded and covered with Portland cement prior to embedment in mortar,

c) The adhesion bond between steel and surrounding mortar can be improved by applying a enamel coating to the steel if the vitreous enamel is coated with or fused to a water-reactive calcium silicate like portland cement,

d) The bond adhesion can be increased by as much as triple that of the uncoated steel,

e) The enamel coating was stripped off of the steel in those samples that showed the strongest adhesion. This suggests that even greater force would be required at pull-out if the bond between the steel and the enamel glass could be increased, and

f) Specially-designed porcelain enamels may have a role in increasing the adhesion bond of concrete to steel as well as controlling the corrosion of the underlying steel.

References

Gray, R. J., and Johnson, C. D. Measurement of fibre-matrix interfacial bond strength in steel fibre-reinforced cementitious composites. Pp. 317-327. Swamy, R. N. 1978. Testing and Test Methods of Fibre Cement Composites. RILEM-The Construction Press. Lancaster, England. 545.

Al Khalaf, M. N., and Page., C. L. 1979. Steel mortar interfaces: microstructural features and mode of failure. Cement and Concrete Res. 197-208.

Bentur, A. Diamond, S., and Mindess, S. 1985. The microstructure of the steel fibre-cement interface. J. Material Sci. 20:3610-3620.

Society of Manufacturing Engineers, (undated). "Porcelain Enameling" Society of Manufacturing Engineers Dearborn, MI

Tattersall, G. H., and Urbanowicz, C. R. 1974. Bond strength in steel-fibre reinforced concrete. Magazine of Concrete Res. 26(87), 105-113.

Acknowledgments: This work was undertaken as part of the Corps of Engineers AT22 Basic Research Program conducted at the U.S. Army Engineer Research and Development Center under Work Package #225, Work Item A011.

USING COEFFICIENT OF THERMAL EXPANSION (CTE) TO HELP SOLVE PORCELAIN ENAMEL STRESS -RELATED PROBLEMS

Andrew F. Gorecki
Ferro Corporation
Cleveland, Ohio USA

Abstract

Results and trends of the thermal expansion and softening points of typical porcelain enamel systems are reported. The technique which combines thermal expansion coefficient testing with industry specific enamel fit testing is promoted to solve stress- related defects seen in porcelain enamel coatings.

Introduction

Enamel users and enamel suppliers discuss the "hardness" of the frit or enamel system when discussing variables relating to porcelain stress related defects. When attempting to describe how "hard" a frit is we often use terms like bond, fusion flow, initial gloss and other industry specific tests. With this paper we would like to promote using thermal expansion when describing porcelain enamel "hardness" and recommend combining coefficient of thermal expansion with industry specific testing as a good approach to solving enamel fit related problems.

Many defects demonstrated in porcelain enameling show root cause with the enamel fit, i.e., the effect of the difference in thermal expansions of the glass and metal substrate. Examples of these porcelain problems include thermal shock in grates, hairline issues in sinks, and crazing of oven bottoms. When attempting to solve stress related porcelain enamel problems industry experts describe the enamel fit through industry specific testing. Here are some of the common fit tests:

- **Warp** [The bending of the enamel-metal composite after firing]
- **Fusion Flow** [Flow of a 2 gram pellet on a vertically positioned enameled sheet, usually 50 mm at near the normal enamel firing temperature]
- **Initial Gloss** [Firing temperature at which the coating becomes glassy in appearance]
- **Hairline Test** [A weighted panel test to promote coating cracking of the bisque]
- **Adherence** [A destructive impact test using a calibrated impact load; example 80 inch-pounds]
- **Thermal Shock** [A water quenching test, usually immersion of a heated panel in room temperature water]
- **Ring Test** [A split ring enameled to determine residual stresses]

Literature examples have attempted to quantify stress in coatings using thermal expansion as a variable. An example of this comes from Kingery[1]. This model for stress is used to determine the stresses in a thin glaze on an infinite slab.

$$\sigma_{gl} = E(T_o - T')(a_{gl} - a_b)(1 - 3j + 6j^2)$$

E= Young's modulus

α_{gl} = thermal expansion glaze

α_b = thermal expansion base

j= ratio of thickness glaze to body

T_o= setting point of glass.

It is important to bring out of this equation the following when talking about stress:

1.) Assuming that the substrate metal stays the same; the expansion coefficient for the glass coating is important.
2.) Assuming that the softening point of a glaze is directly related to the setting point of the glass; the softening point is important.
3.) Assuming that the thickness of the metal stays the same, the coating thickness is important.

The above model, assuming the coating thickness is always the same, proposes that the coefficient of expansion and the softening point and transition temperature dictate the stress seen in the coating. We recommend combining industry specific enamel fit tests with coefficient of thermal expansion and softening point to give the best understanding to solve porcelain enamel fit problems. The following will detail how we perform thermal expansion measurements.

Experimental Procedure

Thermal expansion results were obtained on the Orton Model 1000R dilatometer[2]. The heat up rate was 3°C per minute. The temperature range was room temperature to 700°C. The auto-shut off was used to determine the actual end temperature.

Glass samples for testing were prepared by firing powder or dried slip in a carbon mold. Approximately 8 grams of powder was fired for 12 minutes at 1550°F (621°C). The samples were immediately moved to an annealing furnace set at 1000°F (538°C). The 1000°F furnace was then turned off which allows the samples to slowly cool overnight. The annealing is recommended to avoid breaking the sample during subsequent cutting and polishing. Samples were cut to 2 inches (5.08 cm) using a diamond saw and the ends were slightly rounded using a grinding wheel. Exact length of samples was determined using a micrometer; typical tolerances are +/-10 percent.

Results and Discussion

From the print out of the dilatometer we gather the following data to assist in comparing glasses:

- Linear expansion , Room Temperature (RT) to 300°C(572°F)
  and Room Temperature to 400°C(752°F)
- Temperature of Transition (Tt or Tg)
- Total percent linear change at the softening point

Below are the graphical results from the dilatometer.

Thermal Expansion

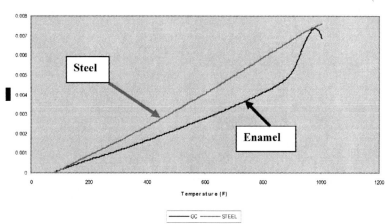

Figure 1 – Coefficients of Thermal Expansion for Steel and a Typical Ground coat Porcelain Enamel

Figure 1 displays a ground coat system and a piece of decarburized steel. The Y axis is the percent linear change and the X axis shows the temperature in degrees F.

Steel has a fairly straight line, as shown in Figure 1, with a constant rate of increase in the size of the piece over the temperature range shown. The glass sample represented by the hooked line has three distinct areas.

The first area of the enamel (or glass) is the relatively straight part going from room temperature until around 850°F (454°C). This area shows that the glass has a constant increase in linear change versus temperature. This is similar to the steel line; however, the rate of increase in length is lower than the steel. In other words, the glass has a lower expansion coefficient.

The second area is called the annealing range. In this area the ground coat shows a change in the rate that the sample increases in length over temperature. For the ground coat shown in figure 1 the annealing range is around 850°F (454°C) until about 975°F (539°C). If you would draw line over the first area and the second area the intersection of both lines would be the glass transition temperature. Usually the glass transition temperature is labeled Tt or Tg. The ground coat in this example has a glass transition of about 850°F.

The third area, located after the annealing range, is the hook. The top part of the hook is called the dilatometric softening point (Ts) for the glass. This is the point where the glass viscosity is so low that the spring loaded dilatometer rod is able to contract the glass. We should note that the actual temperature recorded for the softening point is dependent on the spring used in the dilatometer. The softening point may change from dilatometer to dilatometer and vary over time.

Table I: Product Sort by Softening Pt.

Product	Ts In Degrees C	Ts in Degrees F
ALUMINUM (LEADED)	391	736
GRATE POWDER	494	921
BASECOAT POWDER	507	945
COVERCOAT WET	511	952
CAST IRON GROUNDCOAT	513	955
GRATE WET	518	965
HOT WATER TANK WET	521	970
COVERCOAT POWDER	526	979
BLACK WET	530	986
BLACK POWDER	532	990
COLOR WET	533	991
PYRO WET	533	991
PYRO POWDER	534	993
GROUNDCOAT POWDER	534	933
GROUNDCOAT WET	537	999
HIGH TEMP GC WET	547	1017
CAST IRON COVERCOAT	549	1020

Table I contains experimental results of softening points for a wide range of common porcelain enamel products. We can see that the softening point varies from 391°C (736°F) for a leaded product used over aluminum to 549°C (1020°F) for an enamel cover coat used on cast iron. The normal range for the softening point sheet steel is about 500° to 540°C (932°F to 1020°F).

Table II: Product Sort by Linear Expansion (1E-6in/in C)

Product	Room Temperature to 300°C (572°F)
ALUMINUM (LEADED)	14.93
GRATE POWDER	10.92
BASECOAT POWDER	10.06
COVERCOAT WET	9.60
CAST IRON GROUNDCOAT	9.51
HOT WATER TANK WET	9.46
GRATE WET	9.30
COLOR WET	9.30
CAST IRON COVERCOAT	9.23
GROUNDCOAT POWDER	9.02
BLACK POWDER	8.99
PYRO POWDER	8.92
GROUNDCOAT WET	8.89
HIGH TEMP GC WET	8.73
COVERCOAT POWDER	8.72
BLACK WET	8.66
PYRO WET	8.56

Table II contains experimental results of thermal expansion coefficient from room temperature (RT) to 300°C. The highest expansion is the leaded enamel used over aluminum. It has an expansion of 14.93 x10^{-6}/°C. The sheet steel enamel results are mostly between 8.5 and 10.0 x10^{-6}/ °C. The grate powder is the highest at 10.92 x10^{-6}/°C.

Porcelain Enamel Products

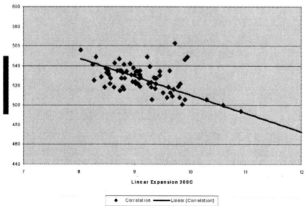

Figure 2 – Softening Pt versus Linear Expansion Values – Room Temperature to 300°C.

Figure 2 is a plot of softening point versus linear expansion up to $300^{\circ}C$. We have drawn a line to show the general correlation between softening point and expansion. As to be expected the lower the softening point the higher the expansion. However, we think it is important to note that that some enamels have very similar expansion coefficient and much different softening points. This gives us the opportunity to choose enamels with different softening points to help solve stress related defects.

General Expansion Trends
- Expansion coefficient of glass lower than metal substrate.
- Heat resistant coatings have higher softening points.
- Good bonding systems have lower softening points.
- Thermal shock resistant systems have higher expansion.
- There is a general trend between CTE and softening point.
- Some enamels have the similar CTE's and different softening points.

We believe that the most success we have had using thermal expansion as a problem solving tool was to use the following format.

- Evaluate expansion of substrate (Metal).
- Evaluate the expansion of the coating layers.
- Correlate with relevant testing such as torsion, hairline, bonding, etc.
- Experiment with coatings with different expansions and/or softening points.

We believe that thermal expansion is a very important variable. However, thermal expansion is not the only variable to consider when we are talking about problem solving stress defects. From previous experience, here is a check list of other variables to look at when investigating stress defects.

- Enamel thickness [face], Enamel thickness[back] and Metal thickness.
- Firing temperature.
- Furnace firing curve [rate of heating and cooling]
- Bubble structure of coating.
- Drying temperature and bisque strength.
- Metal properties, metal and glass reactions.
- Mill addition materials..........Silica, Feldspar, etc.

Summary
Thermal expansion is a good way to help enamel users and enamel suppliers describe frit or enamel system "hardness". We propose that thermal expansion [CTE and softening point] be a tool when attempting to solve stress related porcelain enamel problems.

References
1. W.D. Kingery et al., *Introduction to Ceramics*. John Wiley & Sons, New York, 1976. P 609.
2. The Edward Orton Jr. Ceramic Foundation. Westerville, Ohio.
3. J.Richard Schorr, "Problem Solving with the Dilatometer-Ceramics", The Edward Orton Jr. Ceramic Foundation -Tech Notes Bulletin Number 4.
4. J. R. Taylor and A. C. Bull, *Ceramics Glaze Technology*. Pergamon Press Inc., New York, 1986. P 79.
5. Andrew I. Andrews, *Enamels*. Twin City Printing Co, 1935. P38.
6. Glenn McDonald and Robert C. Hendricks, "Some Thermal Stress Problems in Porcelain Enamel-Coated Rods", Proc. PEI Tech. Forum 42, P. 178(1980).

CHOOSING THE "CORRECT" EQUIPMENT AND ABRASIVE FOR YOUR APPLICATION

Richard Hawkins
Blast-It-All
Salisbury, NC

Abstract
It is hoped that this presentation will benefit you in your future decisions regarding blasting equipment, abrasives, and operator safety.

Systems
Abrasive air blasting can be divided into 2 types. (Suction or Venturi and Pressure). Suction guns are the most widely used our industry primarily because of the maintenance aspect. Pressure Type equipment is considerable faster than suction. Suction equipment uses a venturi effect to pull the media up into the gun from a hopper. The Pressure blasting equipment uses compressed air mixed with abrasive inside a pressure vessel.

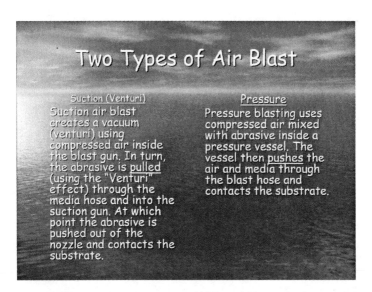

The suction gun is widely used in our industry because of their ease of maintenance, and lower operating costs. But the primary reason is most facilities do not have a compressor large enough to create enough CFM to operate the Pressure Blast Nozzle. I will illustrate the difference a little later in this presentation.

Pressure Air Blasting is usually used in high production applications or when blasting in a blast room, or outdoors. Pressure systems are more maintenance intensive, and require a larger volume of compressed air.

Abrasives
We have more choices than ever for blasting abrasives. You first must determine the results you desire when the blast project is complete. It may be cosmetic, it may be hiding tool marks, it could also include stripping a coating from a substrate, or preparing a substrate for the coating. Each job must be evaluated on a case by case basis. Most of these abrasives can be used in pressure or suction equipment. Some have better recycling characteristics than others. You must always consider the substrate being blasted and the final result desired when choosing the abrasive for your application. I have included Silica Sand on this list only because, it is still widely used by the mom and pop shops, and individual users. Some are not aware of the dangers involved by using this abrasive.

The web address for OSHA regarding the use of silica sand is: www.osha.gov/silica/IT69D_1.html. Please tell anyone you know that long term exposure can cause serious health issues. Even if you are wearing a proper supplied air respirator while blasting, you could still be endangering those downwind of the blast area.

Equipment & recycling
Shown below are various types of blast equipment.

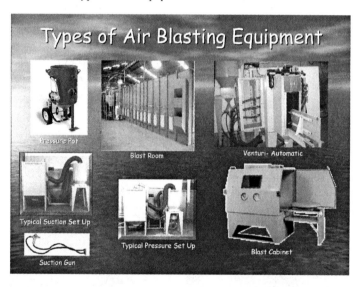

The upper left hand picture shows a standard 6 cu ft pressure pot. This pot can be used as a portable for outdoor blasting or it can be incorporated into a blast room configuration. The upper left hand picture shows an automatic blast cabinet. This cabinet was designed to blast both sides of the aluminum piece shown, using 6 venturi or suction guns. while the piece was automatically conveyed through the blast chamber. The Center picture shows a typical Pressure set up for a cabinet. If you look closely you will se the 1 cu ft pressure pot. Because you are recycling the abrasive through the cabinet, you do not need a large pressure vessel like the 6 cu ft unit. The primary reason for using the larger pot is you do not have stop as often to replenish the abrasive.

When using an abrasive that has good recycling characteristics, you will need to reclaim the blast media for re-use. There are 4 components to consider when choosing your blast equipment. The enclosure isolates the blasting to a dedicated contained area. It could be a cabinet where the operator stand outside the enclosure or a blast room where the operator where the proper safety equipment and enters the blast enclosure. You will need to consider the LARGEST part you will be blasting, and purchase an enclosure large enough to accommodate future projects. The Reclaimer works in conjunction with the Dust Collector to reclaim and separate the dust and debris from the good media. The Blast Nozzle imparts an abrasive and compressed air mixture at about 3 parts compressed air to 1 abrasive ratio. The Dust Collector not only removes the dust from the enclosure but also works with the reclaim to remove the "fines" or dust from the abrasive.

When blasting outdoors it is more difficult to recover the abrasive. so a cheaper abrasive is usually used. When Using a Cabinet or Room to capture the abrasive you must choose the recovery system based upon product being blasted, the abrasive being used, and recovery times required (or Production Requirements). Pneumatics are usually in a blast cabinet or blast room using lighter abrasives and the mechanical recover is used for the heavier abrasives. Of coarse when manually recovering your abrasive it is a broom and a shovel.

Recovery systems
Most pneumatic recovery systems are capable of transporting abrasives weighing from 50# to 130# per cubic foot, considering the floor configurations.

TYPICAL PNUEMATIC FLOOR SECTION

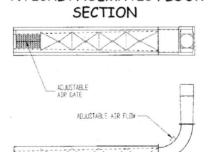

Above is a top and side view of a Standard pneumatic "M" floor section. The abrasive is swept into the floor section by the operator and it is moved pneumatically through the ductwork and through the reclaimer for recycling.

Shown below is a top view of a "U" shaped floor system.

"U" Shaped Recovery Floor

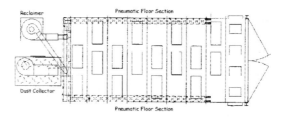

The abrasive is swept back to the floor section which can be located in a pit or above the work floor. The cross floor is known as the transition and it carries the abrasive to a central duct work assembly to be transported back to the reclaim. Also shown in this drawing is the reclaim, and dust collector.

For those whose capital expenditure budget is a little smaller we have the single floor recovery type (see below). You will sweep twice as far and probably take twice as long to get the abrasive back to the pressure pot to begin the blasting process again.

Single Floor Section Recovery System

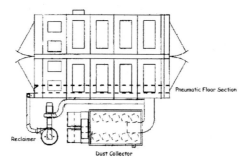

Finally, the Single Point Recovery is for the very budget minded.

Single Point Recovery

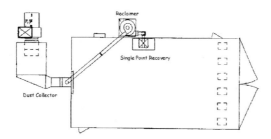

You will spend as much time sweeping as you will blasting with this type of recovery system.

Believe it or not, this type of system was constructed for a very large Aerospace manufacturer. Their production rates were very low so they did not require a fast recovery time, so they opted for this type of equipment. This is the single point recovery, and it was mounted on the wall about 18" off of the ground which meant there was a "shovel" required as well as a broom. Thankfully it was for a plastic media room, and plastic weighs around 50# cu ft.

Single Point Recovery "Detail"

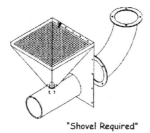

"Shovel Required"

Comparisons

Pneumatic Systems are always less expensive to purchase and operate versus Mechanical Systems. Budgets and Production requirements always dictate the type of recovery system that is right for you.

Mechanical systems typically use Screws or augers to move the abrasive back to a bucket elevator system for removal of the debris and dust. Very seldom will you see a mechanical recovery floor section use any abrasive other than steel grit or shot.

Mechanical "U" Recovery

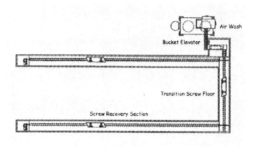

"U" Shaped Recovery Floor

This type of system works very well for heavier abrasives such as those mention here. Taking into consideration the weight of the abrasive, this type of floor configuration offers the least sweeping with the most economically minded budget. If you are required to use steel grit or shot.

This configuration offers a higher production rate because there is less sweeping. On occasion it is facility space constraints that determine the "H" shape rather than the "U" shape recovery system. In other words the plant does not have room at the end of the blast enclosure so they place the elevator and dust collector in the center.

Mechanical "H" Recovery

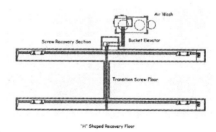

"H" Shaped Recovery Floor

Below is a single point recovery system using a 3' X 3' profile to capture the abrasives. This systems floor section is located in a concrete pit and all the abrasives are swept to this area. This is the least expensive, and the only screw or auger in the system is the cross above the bucket elevator going over to the scalper screen.

Mechanical "Single" Point

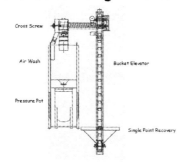

Standards

Once you have determined your blast enclosure, recovery system. and production requirements, you will need to determine the blast standards or how clean do you want your product. The Society for protective coatings has established standard for the coatings industry. They are Brush Off, Commercial, Near White, and White Metal. Brush Off or commonly known as SP-7 when viewed without magnification, may only have tightly adhering residues of mill scale, rust, and coatings remaining and must have evenly distributed flecks of metals exposed. Commercial or SP6 when viewed without magnification, must have at least two-thirds free of visible residue. For most applications where standard coatings will be applied, commercial blast is specified.

With Near White finish or SP 10 when viewed without magnification, at least 95% of the surface must be free of visible residue. Similar to WHITE METAL but slight staining is allowed. This is usually used in applications where High Performance Coatings are being applied. White Metal or SP-5 when viewed without magnification shows a White Metal surface is free of all visible rust, scale, paint and foreign matter. Usually used where Zinc Rich Coatings are being applied.

DEGREES OF CLEANLINESS

Key - Steel where mill scale has started to flake and light rusting occurs.

Key - Steel where all mill scale has flaked off and Rusting has taken place.

Key - Steel where pitting and complete rusting has occured

Original Condition of Steel before Blast Cleaning

Brush Off Finish

Commercial Finish

Near White Finish

White Finish

The column to the left- top picture shows the steel has mill scale which has started to flake and light rusting has occurred. The center picture on top shows the mill scale has flaked off rust has taken a good hold, and finally, the top picture on the right shows the Mill scale is gone and severe rusting and pitting has occurred. The other pictures show the conditions after blast cleaning to Brush Off, Commercial, Near White, and White Metal.

The web addresses for the Society of Protective Coatings and the National Association of Corrosion Engineers are www.sspc.org and www.nace.org. These societies and their websites are a wealth of information.

Choosing your abrasive
Abrasives can be classified into one of three categories - Natural, Manufactured, and By-Products. Natural abrasives are those that are mined or are "from the earth," the manufactured abrasives are those which are designed for the abrasive blasting market. They are designed to give a certain cosmetic finish or profile. And finally the by- Products are those where they have been discovered to have good blasting characteristics but were not originally intended for the abrasive industry. A profile is the appearance of the peaks and valleys of a finished abrasive blast cleaned substrate. When cosmetically finishing a part you see a consistent blast pattern across the surface, and usually no coating will be applied after the blast process. The factors to consider when choosing your abrasive are size, shape, density, hardness, friability.

Abrasives are classified using a sieve that has been certified to a mesh size. They are then manufactured into a mesh range and packaged for shipment. Some abrasives use the range as their designation such as a 30/60 coal slag or a 12/16 plastic. Other abrasives are designated as a Letter or a number or a combination of both. Such as G8 glass bead which has a mesh range of 70-100.

Abrasives are either round or angular. Both impart different results on the substrate. The result you desire will determine the shape of the media you use.

Not only does the density of the media determine the type of recovery system you will use, it also influences the finished product being blasted, whether it be by peening the substrate, or creating a profile for a coating application.

The hardness of abrasives with the exception of steel grit & shot are measured using the MOH's Scale. With 1 being the softest and 15 being the hardness. Diamond measures in a 15 on the MOH's scale. The more delicate the substrate the softer the abrasive needs to be.

Friability refers to the abrasives ability to be cycled through the nozzle. The harder an abrasive is the longer it will last, and ultimately costing you less operating expense. Sand will typically only last (1) cycle, steel grit and shot up to 200 cycles.

The below chart shows the mesh size, shape, density, hardness, recyclability, cost, and source.

ABRASIVE COMPARISON CHART

Material	Mesh Size	Shape	Density Lbs/ cu. ft.	Mohs Scale	Friability	Cost	# of Cycles	Source	Application
Silica Sand	6-270	x/o	100	5.0-6.0	high	low	1	Natural	Outdoor Blasting
Slag	8-80	x	85-112	7.0-7.5	high	med	1-2	By Product	Outdoor Blasting
Walnut Shell	6-100	x	55	5.0	high	med	1	Natural	Cleaning, Polishing
Glass Bead	10-400	o	85-90	5.5	medium	med	8-10	Mfg	Cleaning, Finishing
Aluminum Oxide	12-325	x	125	8.0-9.0	medium	high	15-20	Mfg	Cleaning, Deburring, Etching
Plastic	10-80	x	45-60	3.0-4.0	medium	high	8-10	Mfg	Stripping, Deburring
Steel Grit	10-325	x	230	8.0	low	high	200	Mfg	Removing
Steel Shot	8-200	o	280	8.0	low	high	200	Mfg	Peening, Cleaning
Crushed Glass	12-400	x	110	5.0-6.0	med	med	3	By Product	Cleaning, Etching
Garnet	10-300	x	120	7.0-8.0	high	med	5	Natural	Cleaning, Etching

x- Angular Abrasive
o- Round Abrasive

Natural- Naturally occurring abrasive
By- Product- Result of another Process
Mfg- Manufactured for abrasive blasting

The below chart shows the Anchor Pattern or Profile for each abrasive.

Anchor Pattern or Profile

ANCHOR PATTERN

Abrasive Type	Abrasive mesh size required to produce this depth anchor pattern						
	0.5 mil	1 mil	1.5 mils	2 mils	2.5 mils	3 mils	4 mils
Garnet	100	80	60	40	36	24	16
Coal Slag	-----	-----	-----	30/60	20/40	16/30	12/20
ALOX	120	80	60	40	36	24	16
Walnut Shell	-----	20/30	12/20	-----	-----	-----	-----
Steel Grit	200	120	80	50	40	25	16

Typically Anchor Patterns are expressed in Mils, Microns, or Millimeters.
1 Mil= 1/1000 inch
25 microns= 1 Mil
39.4 Mils= 1 Millimeter

Plastic Media will generally NOT create a profile on a typical steel substrate.

Glass Bead will NOT create a significant profile

Steel Shot is used for stress relieving or Shot Peening.

The chart below is based upon an abrasive that weighs approx 100/ per cu ft.

NOZZLE & COMPRESSED AIR CHART

Nozzle Size	Pressure @ the Nozzle (psi)							
	50	60	70	80	90	100	125	
No 2 1/8"	11	13	15	17	18.5	20	25	Air (CFM)
	67	77	88	101	112	123	152	Lbs. / Hour
	2.5	3	3.5	4	4.5	5	5.5	Compressor HP
No. 3 3/16"	26	30	33	38	41	45	55	Air (CFM)
	150	171	196	216	238	264	319	Lbs. / Hour
	6	7	8	9	10	10	12	Compressor HP
No. 4 1/4"	47	54	61	68	74	81	98	Air (CFM)
	268	312	354	408	448	494	608	Lbs. / Hour
	11	12	14	16	17	18	22	Compressor HP
No. 5 5/16"	77	89	101	113	126	137	168	Air (CFM)
	468	534	604	672	740	812	982	Lbs. / Hour
	18	20	23	26	28	31	37	Compressor HP
No.6 3/8"	108	126	143	161	173	196	237	Air (CFM)
	668	764	864	960	1052	1152	1393	Lbs. / Hour
	24	28	32	36	39	44	52	Compressor HP
No. 7 7/16"	147	170	194	217	240	254	314	Air (CFM)
	896	1032	1176	1312	1448	1584	1931	Lbs. / Hour
	33	38	44	49	54	57	69	Compressor HP
No 8 1/2"	195	224	252	280	309	338	409	Air (CFM)
	1160	1336	1512	1680	1856	2024	2459	Lbs. / Hour
	44	50	56	63	69	75	90	Compressor HP

Abrasive Consumption based upon an abrasive that weighs 100 pounds per cubic foot. When sizing compressor add 50% to HP to allow for "wear" on nozzle.

The BLUE numbers Column show all of the specs when blasting @ 80 PSI. The RED numbers show @ 125 PSI. We recommend blasting @ 80 PSI or Less. When you blasting @ higher than

80 PSI you will usually NOT get any faster production, but you WILL BREAK the abrasive down at a faster rate resulting in higher operating cost. If you require blasting @ higher than 80-90 PSI you should look at going to a more aggressive mesh size or changing your abrasive type all together.

GRIT BLASTING CASE STUDY

Liam O'Byrne
O'Byrne Consulting, Kingsburg, CA.

Abstract
This paper briefly describes a project requested by HKF Industries, of Los Angeles, CA for an analysis of blasting processes and controls at their plant located in Fuzhou, Fujian Province, China.

The project was designed to initially audit the overall blasting process and process controls, recommend possible improvements for the process, and then implement and measure gains from the improvement opportunities identified.

The overall goals of the project were to obtain Quality, Productivity and Process Consistency improvements.

Initial Audit Findings
During the initial process audit, the following were identified as findings of fact, upon which to base possible improvements:
- One 4-wheeled spinner/hanger blaster was performing all blast cleaning operations.
- No annealing prior to blasting was being performed.
- Abrasive grit blasting media was too soft for the process, and showed rounding tendencies after several cycles in the machine.
- The average total blast cycle time was between 20-30 min, depending on the part being blasted.
- There was a heavy part density on the hangers in the machine, and this was causing interference and "masking" from part to part.
- The castings were being moved around on the hanger arms by the force of the blast stream. This was due to a lack of positive locators on the hanger arms.
- The abrasive blast media was not monitored on a consistent basis for size distribution.
- The amperage pulled by the blast wheel motors varied over the course of a shift, signifying variations in blasting conditions in the machine and cleaning efficiency.
- The overall machine condition showed signs of needing more continuous maintenance attention.
- There was a high incidence of castings needing to be processed with a two coat/two fire operation, even on darker colors.

Project Outline
Based on the above findings, a project outline was developed with the following key areas identified for particular attention:
- A harder abrasive grit media would be immediately introduced to provide a better etch on the casting surfaces.

- The hanging density of parts would be adjusted to provide better impact of the abrasive grit media on the parts.
- The overall blasting process would be adjusted for more efficient coverage and shorter blasting time.
- A preventive maintenance system would be implemented for maintaining consistent blasting efficiency and improved machine maintenance.
- An overall blasting process would be designed to achieve one coat/one fire enamel coating on dark colors as a minimum goal.

Project Implementation

The following items were implemented as a means of achieving the goals in the project outline:

- The abrasive grit media was increased in hardness from 42-50HRC to 50-55HRC as a minimum.
- The part density on the hangers was reduced by 20%.
- Positive locators were added to the hanger arms to prevent the movement of parts in the blast stream.
- A system of consistent additions of new grit was implemented during the work shift, to maintain grit size distribution in the blaster.
- A process for sieve analysis of the grit size distribution was implemented on each work shift.
- Blast pattern checks were implemented on a weekly basis for each blast wheel.
- General daily/weekly preventive maintenance checks on the blast machine were implemented, with special attention being given to the blast wheels.
- Regular monitoring of the wheel motor amperage was instituted as a process for maintenance control.
- Blast timing cycles were checked for each specific part and reductions in overall cycle times were introduced as appropriate.
- A one coat/one fire enamel process was tested in the enamel department once the above changes were implemented.

Project Results

The following results from the changes implemented were obtained:

- The total blast cycle time has been reduced by 50+% for all castings.
- A more consistent and efficient cleaning process has been demonstrated using etch profile measurement.
- One coat/one fire enamel coating is now a standard practice for all dark colored enamels, still without annealing any castings.
- The overall enamel yield rates historically experienced have been maintained when measured against the two coat/two fire process.
- The resulting enamel productivity improvements have led to the purchase of new blast machines to match the improved overall production capacity in enameling.
- The new blast machines have been commissioned using the same process controls as developed for the original blaster.

PHOSPHATE REPLACEMENT TECHNOLOGY

Ken Kaluzny
Coral Chemical Company

Abstract
Phosphating of steel and aluminum substrates is commonly associated with industrial painting processes. However, phosphate conversion coatings create effluents that are environmentally unfriendly. Development of a vanadium conversion coating is described along with a case study indicating lower energy and wastage resulting in a high return on investment with the new process. The vanadium conversion coating was shown to have enhanced corrosion resistance with a variety of organic coatings on salt spray exposure.

Fluorozirconium coating technology

> inorganic conversion coating

> complexed transition metal salts (i.e., vanadium)

> thin film similar to a traditional chemical conversion coating

> replace conventional iron phosphating products and some zinc phosphate

3

Benefit of VCCs

- Reduced/Eliminate Phosphate Discharge
- Ambient Temperature
- Mild Steel Equipment
- Reduced Maintenance
- Reduced Waste Water Treatment
- Reduced Rinse Water Use
- Lower costs

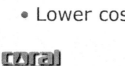

History of Vanadium

- Vanadium Is A Transition Metal Used Commonly To Improve Steel
- Used For Producing Tools, Rust-Resistant Springs, Etc.
- Use in chemical formulations as a replacement for chromate based treatments for aluminum alloys for painted and non painted applications.

History of VCCs

- Vanadium compounds enhance the corrosion protection of anodized coatings and are corrosion inhibitors for Al alloys.
- VCCs are patented for use on magnesium alloy substrates in preparation for painting.
- In 1999, Coral Chemical patented a vanadium based final seal rinse product. A vanadium treatment for steel substrate introduced in 2005 "patent pending".

Sources: H. Guan, R.G. Buchheit, Fontana Corrosion Center, Dept of Material Science & Engineering, Ohio State University, Columbus, Ohio

Narayan Das, John P. Jandrists, "Non-Chrome Rinse for Phosphate Coated Ferrous Metals," U.S. Patent no. 6,027,579 (2000).

Characterization of VCCs

Vanadium Reacts With Oxygen To Form A Vanadium (V) Oxide Conversion Coating, V2O5

$4V(2) + 5O2(G) \rightarrow 2V2O5(S)$ [yellow-orange]

Characterization of VCCs

The process creates an amorphous surface film comprised of hydrated vanadium oxides.

An inorganic polymerization of V5+ is created which leads to the buildup of a film that passivates the surface and inhibits corrosion.

Under x-ray absorption testing, vanadium oxides and other components can be detected on ferrous and non-ferrous substrates.

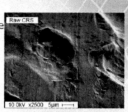

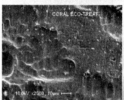

Temperature Reduction Benefits

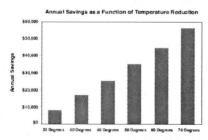

Energy savings is based on a 2,000-gallon tank operated 16 hours a day, 50 weeks a year with a natural gas cost of $6/mm BTU.

      Let Us Take Your Temperature

VCC vs Iron Phosphate salt spray results

Paint	VCC	Iron Phosphate	Hours	Substrate
Polyester Powder	0.3 mm	4.3 mm	504	CRS
Hybrid Powder	2.5 mm	4+ mm	504*	CRS
Cathodic E-Coat	0.6 mm	3.2 mm	504	CRS
Cathodic E-Coat	0.2 mm	6.3 mm	1,000	CRS

* vanadate coating achieved 888 hours

11

Salt Spray Performance

Substrate	Paint	Eco-Product	Eco-Performance	Iron Phos
CRS	Polyester Powder	Eco-Treat	0.3 mm	4.3 mm
CRS	Polyester Powder	Eco-Treat DP	0.6 mm	4.3 mm
CRS	Hybrid Powder	Eco-Treat	2.5 mm	4.0 mm
CRS	Cathodic E-Coat	Eco-Treat	0.6 mm	3.2 mm
CRS	Cathodic E-Coat	Eco-Treat DP	1.4 mm	3.2 mm
CRS	Hybrid Powder	Eco-Treat NP	3.0 mm	3.0 mm
Alum	Polyester Powder	Eco-Treat NP	0.0 mm	N/A
Alum	Polyester Powder	Eco-Treat DP	0.0 mm	N/A
E-Galv	Cathodic E-Coat	Eco-Treat	2.5 mm	N/A
CRS	Epoxy Wet Spray	Eco-Treat	0.3 mm	0.8 mm
HRS	Epoxy Wet Spray	Eco-Treat	0.1 mm	0.4 mm

Product	Washer Size	Stage Used	Substrate	
Eco Treat	3, 4, 5	3, 4, 5	Steel and Galvanized	**500 SS Hours**
Eco Treat NP	3, 4, 5	3, 4, 5	Aluminum and Steel	**ASTM B117**
Eco Treat DP	3 +	Final	Aluminum and Steel	

Product Types

Phosphate Free Cleaners
90 F Operating Temp
Light to Moderate Soils

Low Phosphate Treatment
Very Low Phosphate
Ambient Temperature
Outstanding Performance
Vanadium Oxide Technology

No Phosphate Treatment
Phosphate Free
Ambient Temperature
Transitional Metal Technology

"Dry-In-Place" Treatment
Phosphate Free
Ambient Temperature

VCC Final Rinses
Phosphate Free
Ambient Temperature
Low Concentration

coral

ROI Analysis for Vanadate Conversion Coating

- California Company – Annualized Gain

Savings
    Nozzle/riser maintenance $1920
    Hauling Costs $1200
    Energy savings in Gas $9600
        Totals $12,720
Costs
    Increased product usage   ($3600)
    Increased product cost   ($1200)
        Net Investment   ($4,800)

14   *First year net return = $7,920*
ROI= 165%

CASE STUDY – PROCESS STABILIZATION BY CONTINUOUS OIL CONTAMINANT REMOVAL

Jim Polizzi
KMI Systems

David Latimer
Whirlpool Corporation

Jack Scambos
ARR, Inc.

Abstract
Oil is universally used as a lubricant in the enameling industry in the metal forming sections of the manufacturing operations. This oil must be removed before subsequent processing, particularly if an aqueous based enamel system is applied to the metal.

A new type of mechanical oil separation technology is described known as the "Suparator". This new methodology takes advantage of small differences in hydrodynamic pressures as liquid flows over a series of baffles near the surface of the liquid layer that has oil floating on it. The oil accumulates as the water and cleaning agents are displaced and then is collected for removal by a "waterfall" arrangement.

The new system has been shown to extend the life of cleaning solutions and reduce the wastage of cleaning products. Several cases of improved performance as cited. A wide variety of industries are using this technology with over 5500 units in operation worldwide at this time.

What are some questions we should be asking ourselves?

- What is our #1 source of defects (excluding the coating itself)?

- What is the #1 method we all use to clean our parts?

- What is the #1 source of cleaning failure?

- What is the obvious solution to the aforementioned problem?

- What is the most difficult cleaner contaminant to control?

- What do we need to do to eliminate this problem?

What exactly, is oil/water separation?

- The process, usually mechanical, of removing insoluble oils (and oil-based contaminants) from a water source such as cleaner baths, and wastewater discharge flows and sumps.

Why is this important to our industry?

- Excessive oil in the cleaner bath = poor adhesion = defects = increased costs
- Excessive oil in the cleaner bath = $$$
- Increased wastewater treatment cost and environmental issues
- Increase maintenance activities + decreased operating efficiency
- Regulatory standards
- Competitive advantage

What should we do?

- Controlling oil contamination in our finishing process is one of the easiest, yet extremely effective things we can do to reign in costs

  1. Keeping the cleaner "clean" will result in longer periods of time between dumping of the baths – i.e.: longer cleaner life = less chemicals required = lower operating costs

  2. Keeping the cleaner "clean" will result in less effluent being sent to your wastewater treatment facility + what is sent is easier to treat = lower overall operating costs.

  3. Removing oils from your system will result in less maintenance and less downtime due to fouled equipment = lower overall operating costs and frees up your maintenance staff to address other issues in your facility.

  4. Any contaminant that can be removed from your process makes it that much easier to be compliant with regulatory standards. The benefits of this are obvious...

To re-cap, removing oil results in the following

1. A better finished product (fewer rejects)
2. Improved manufacturing process (higher efficiency)
3. Better Process Control: i.e. Repeatability
4. Improved water treatment
5. Environmental compliance
6. Lowered cost per unit

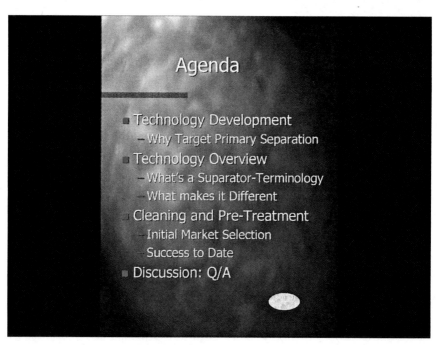

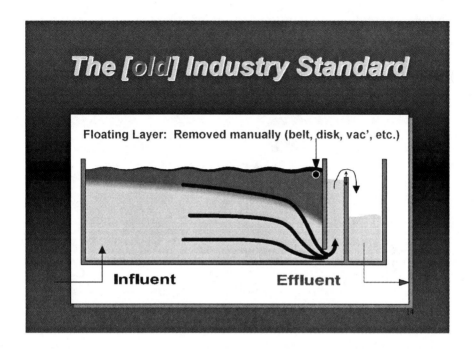

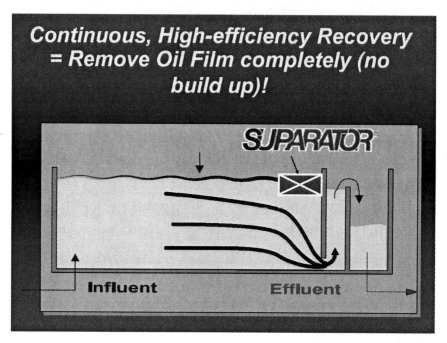

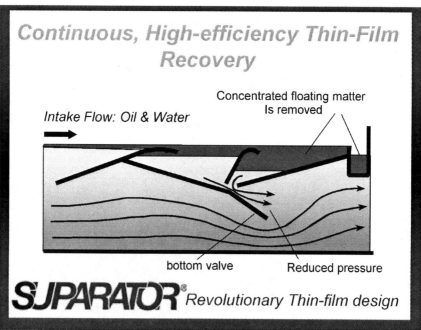

Step-by-Step

1. First the upper fraction of the flow is separated from the rest, carrying along any trace of oil.
2. Next all oil traces are collected and accumulated while water is drained through the bottom section.
3. Oil is concentrated in a small area; remaining water and surfactants are displaced from the concentrate.
4. Finally pure oil is separated

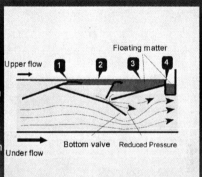

18

Oil Removal - The Trap

Interface creation is proportional to contact time

- *Adjustments to water level set interface elevation*
  At same time, oil density and qty. effect it:
  Can never find interface!!

Two choices for System Operation:
- *Thick oil layer: penalty of large interface layer*

- *Thin Oil Layer: Remove quickly but lots of water*

Engineered-for-Aqueous Cleaning

19

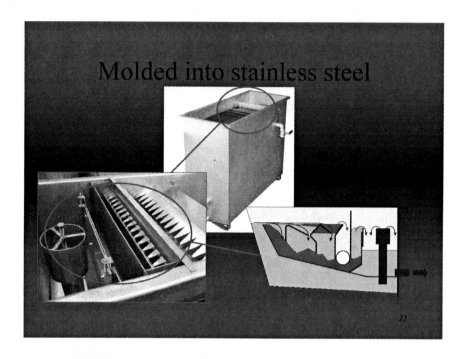

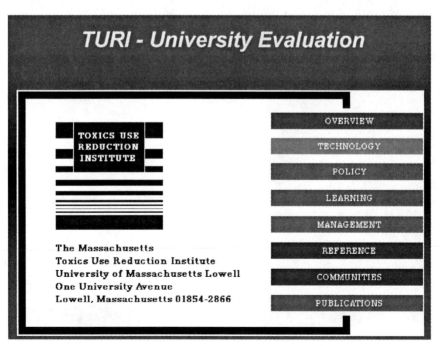

TURI Selected Sampling of Cleaning Processes

Case Study #1 **International Automotive Parts Manufacturer** *Alkaline Detergent*	**Case Study #2** **Manufacturer of Premiere Cookware** *Neutral pH (SBN)*
Case Study #3 **"Job Shop" Plating - Electroplating Specialty** *Oil-ejecting Alkaline*	**Case Study #4** **Bulk Heat Treater - In-Plant Operations** *Hot Water only*

26

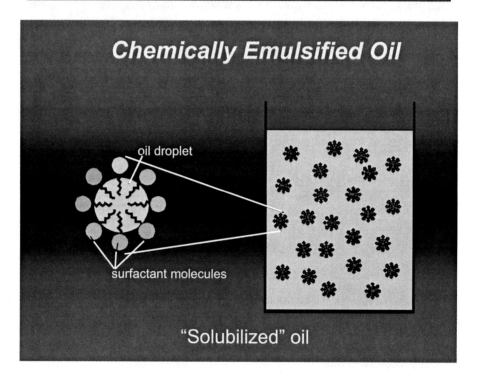

Chemically Emulsified Oil

oil droplet

surfactant molecules

"Solubilized" oil

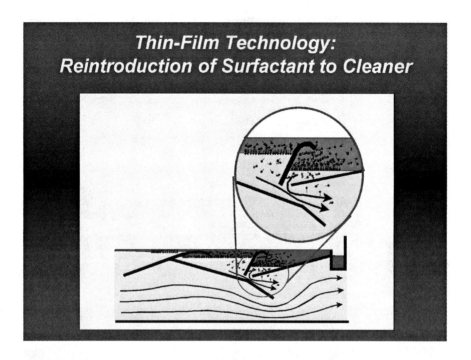

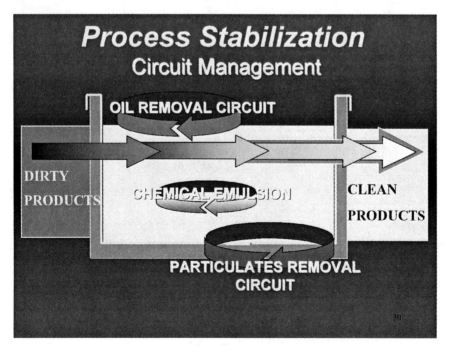

What's Being Said:

➤ *"Suparator will save us $101,184 in the 12 months since it's installation in disposal and downtime alone"*
Parker Hannafin Hydraulic Hose Division

➤ *"The improvement in our pre-treatment quality has virtually eliminated what was a persistent finish quality issue"*
HON / Allsteel Industries

➤ *"We no longer have to ship a particular class of our parts outside to a contract cleaner"*
Daiki, supplier to Komatsu and Kubota

➤ *"Suparator reduced our brazing rejects 99%"*
Delphi Fuel Systems

➤ *"The savings we achieved on our first Suparator allowed us to purchase five additional Suparator Systems"*
Haworth Furniture Systems

Engineered-for-Aqueous Cleaning

31

Installed Base

-Over 5500 units on-line in USA:

- **Parts Washers (90+ %)**

- **Machining Coolant**

- **Paint Pre-Treatment**

- **Heat Treat lines (120+)**

-We maintain the largest Knowledgebase of fluid maintenance applications in the World

Engineered-for-Aqueous Cleaning

32

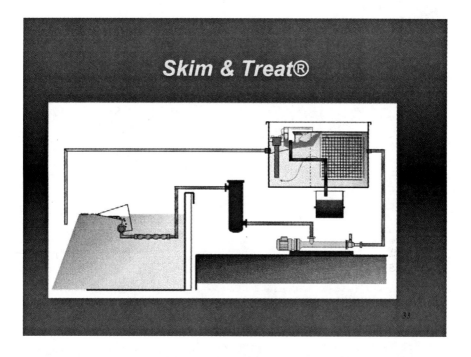

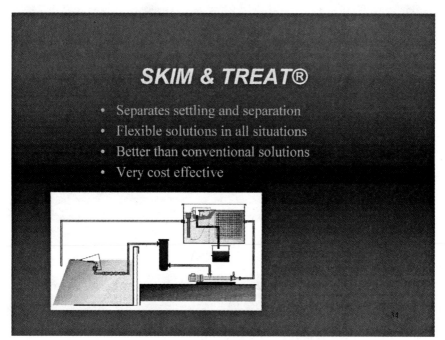

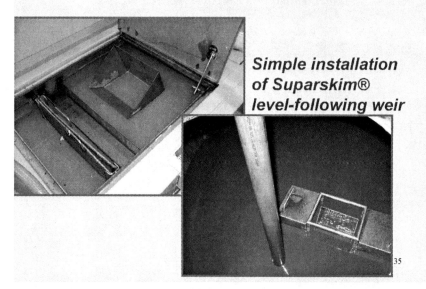

Configuring Your System

Simple installation of Suparskim® level-following weir

Industry Recognition

- Specified in new DCX, Ford and GM Powertrain Programs

- Selected for Henry Ford Award for Manufacturing Technology Excellence

- EPA STEP Program Technology Evaluation – Full Report available

- Created ASME National manuf. Week Roundtable discussing concept of "Second Generation AQ Cleaning"

-Member of FoMoCo "Fill for Life" Program developed to reduce contamination issues at all automatic transmission plants

Engineered-for-Aqueous Cleaning

Appliance Industry

Competitiveness with an AUP of $13-17K

- *Immediate cost savings in downtime, waste treatment, energy and consumables*

- *New lubricants require better, more complete, oil removal*

- *Steel Sheet stock arriving with a wider variety of oils on them – cost of petroleum*

- *Process stabilization is the key to predictable quality and minimal waste and rework*

- *New cleaner formulations designed to be completely recyclable, less chemical emulsification of insoluble oils and less saponification of fats (these processes consume cleaner components and limit useable cleaner life).*

Engineered-for-Aqueous Cleaning

Conclusions

- *Classic Oily~Water Separators count on long, static quiet times and the build-up of thick oil layers to gain efficiency = long contact time & S.A. = maximum interaction & contamination*

- *Suparator's Dynamic Thin-Film Technology recovers the oil In Flow prior to and without the requirement of thick layer build-up = less contact time = less interaction*

- *Oily Solids and associated materials are contaminants too and should be continuously removed, to the greatest extent possible, in real-time with the oil – not ignored / settled for manual removal from the bottom of a settling tank or column.*

- *We have a proven track record of changing industry prospective regarding aqueous fluids' useable life, disposal, and performance.*

Engineered-for-Aqueous Cleaning

Results To date

- Less variation in measured process parameters

  - Total Available Alkalinity
  - Total Solution Conductivity

41

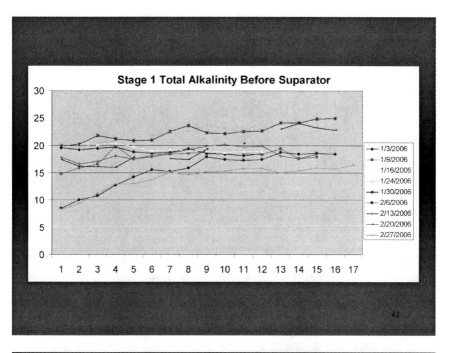

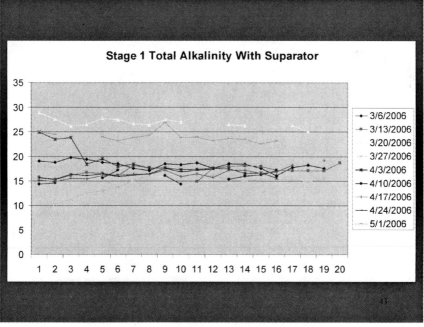

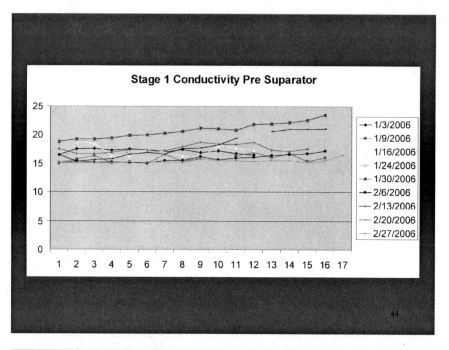

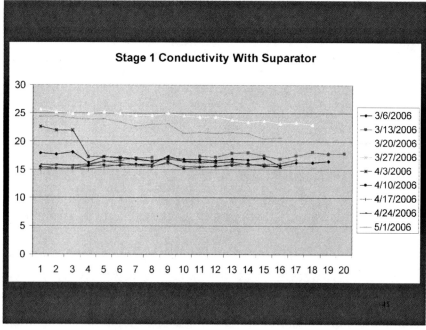

LIGHT EMITTING CERAMIC DEVICES (LECD)

Richard Begley
Marshall University Research Corporation

Don Osborne
Ecer Technologies

Abstract
The light emitting ceramic device (LECD) is a patented technology which is used to produce light by the controlled discharge of electrons through special ceramic compounds. Electrons from current flow excite the light phosphors which causes them to emit photons. A metal core of 22 gage stainless steel is enameled to provide a coating with specific dielectric properties. Subsequent layers of ceramic materials are used to develop the light emitting device. These devices are sealed in a ceramic coating (porcelain enamel layer) which results in a very long life light generating device. The resultant devices are very energy efficient regarding the conversion of electric energy to light with exceptionally low heat generation.

The Technology

LECD (Light Emitting Ceramic Device)
The LECD is patented, exciting new lighting technology. Light is produced by creating a device which controls the discharge of electrons through special ceramic compounds. The electrons excite special light phosphors causing them to emit photons. The resulting reaction is very efficient with the bulk of the resulting energy conversion being light. This means NO HEAT. Since the light phosphors are sealed in ceramics, the lamp has a very long life.

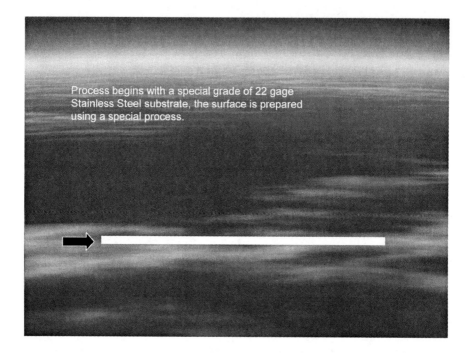

Process begins with a special grade of 22 gage Stainless Steel substrate, the surface is prepared using a special process.

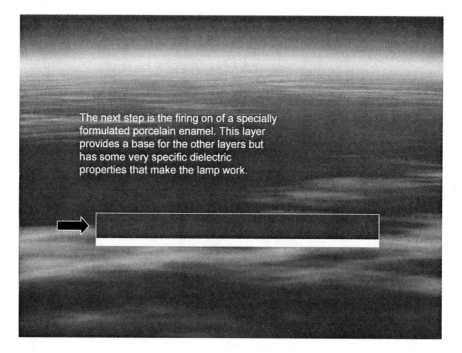

The next step is the firing on of a specially formulated porcelain enamel. This layer provides a base for the other layers but has some very specific dielectric properties that make the lamp work.

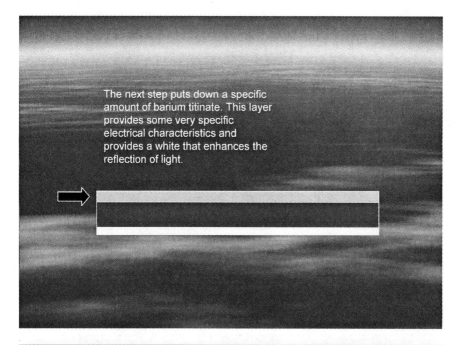

The next step puts down a specific amount of barium titinate. This layer provides some very specific electrical characteristics and provides a white that enhances the reflection of light.

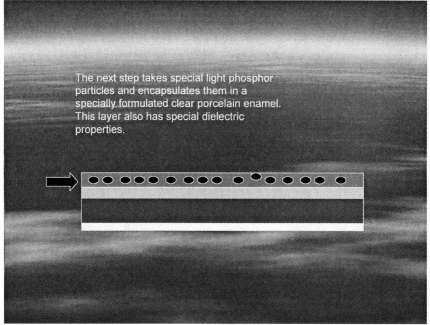

The next step takes special light phosphor particles and encapsulates them in a specially formulated clear porcelain enamel. This layer also has special dielectric properties.

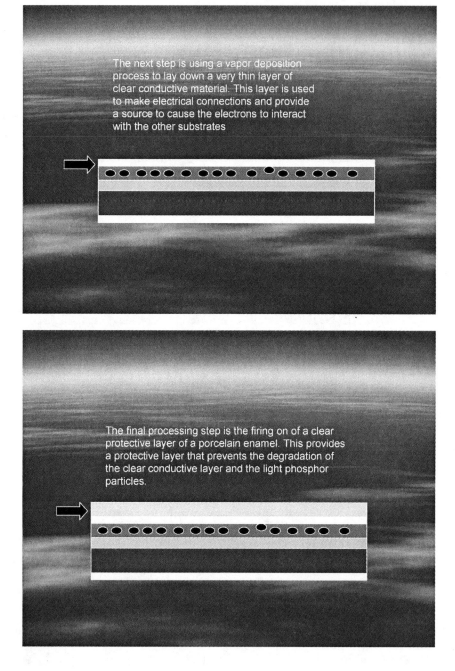

The next step is using a vapor deposition process to lay down a very thin layer of clear conductive material. This layer is used to make electrical connections and provide a source to cause the electrons to interact with the other substrates

The final processing step is the firing on of a clear protective layer of a porcelain enamel. This provides a protective layer that prevents the degradation of the clear conductive layer and the light phosphor particles.

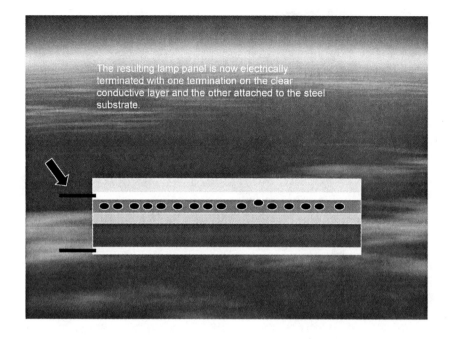

The resulting lamp panel is now electrically terminated with one termination on the clear conductive layer and the other attached to the steel substrate.

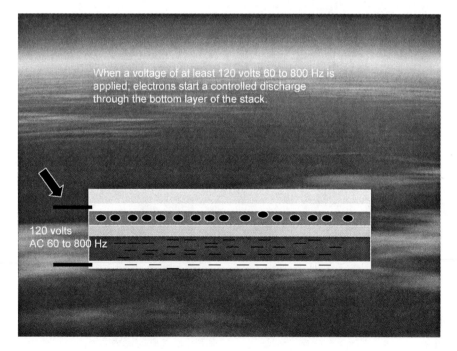

When a voltage of at least 120 volts 60 to 800 Hz is applied; electrons start a controlled discharge through the bottom layer of the stack.

120 volts
AC 60 to 800 Hz

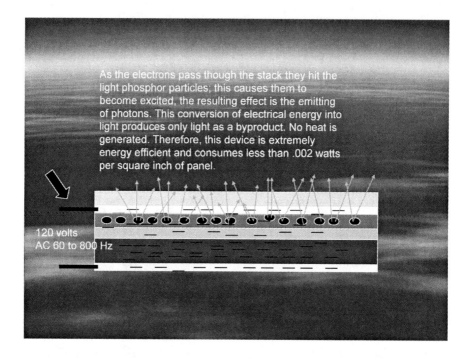

As the electrons pass though the stack they hit the light phosphor particles; this causes them to become excited, the resulting effect is the emitting of photons. This conversion of electrical energy into light produces only light as a byproduct. No heat is generated. Therefore, this device is extremely energy efficient and consumes less than .002 watts per square inch of panel.

120 volts
AC 60 to 800 Hz

Partners In Development

- US Department of Energy (DOE)
- Marshall University
- Rahall Transportation Institute
- West Virginia Economic Development Office
- Osram Sylvania
- Alfred State University
- AO Smith Corporation
- Meadow River Enterprises, Inc.

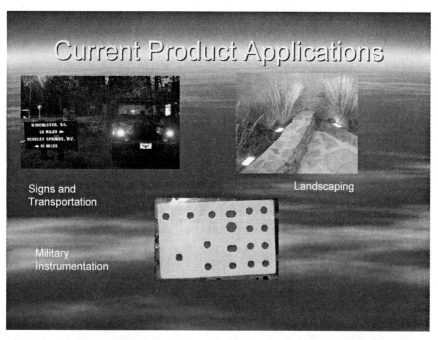

ENAMELS, COLOR MATCHING MODEL*

M. Leaveaux, V. Duchamp, M. F. Perrin, and D. Patou
Pemco Corporation

Abstract

As for a great number of coatings applied on a substrate, enamel also has to be colored besides its resistance and durability qualities. In applications such as architecture, culinary articles, domestic or sanitary appliances, the visual aspect is important as well as all other more technical characteristics. As a result, the professional in the field–enamel supplier or enameller, is continuously confronted with new requests from designers or marketing people. Today these requests have to be satisfied in an increasingly short period.

If in the past the color was especially a matter of practical experience, the art man having to acquire as much knowledge as possible about frits, pigments and their combinations, today we cannot content ourselves with this situation and this for flexibility, reactivity or capacity reasons.

Computer science can reinforce and complete this experience, and at the same time contribute to accelerating the whole color matching process. In this presentation, two different ways are presented. The first one uses a preliminary studied mixture plan, which enables, by using simple mathematical laws, approaching the searched result on condition that it is possible to realize this result starting from this preliminary set mixture plan.

The second one examines the pigment characterization, which in this way creates a database used by calculation software that recomposes the required color. This system, in comparison to the previous one, has no other limit than the art limit. The two systems that were initially used in other industries (plastics, paintings, inks) have been adapted to the enamel industry.

---

*Presentation given by Steve Kilczewski

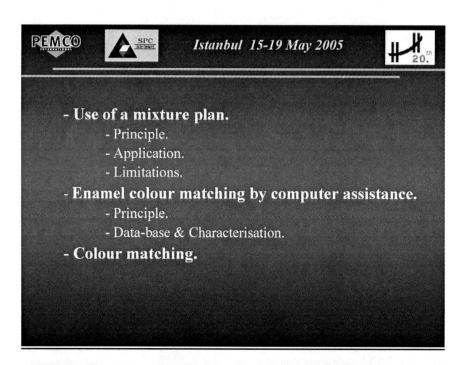

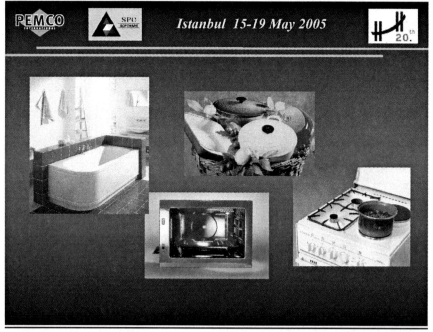

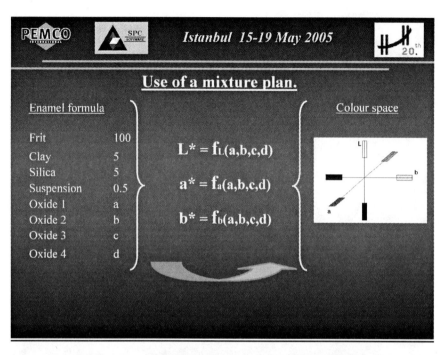

PEMCO | SPC SOFTWARE | *Istanbul 15-19 May 2005* | 20th

Use of a mixture plan.

Enamel formula

Frit	100
Clay	5
Silica	5
Suspension	0.5
Oxide 1	a
Oxide 2	b
Oxide 3	c
Oxide 4	d

$$L^* = f_L(a,b,c,d)$$

$$a^* = f_a(a,b,c,d)$$

$$b^* = f_b(a,b,c,d)$$

Colour space

PEMCO | SPC SOFTWARE | *Istanbul 15-19 May 2005* | 20th

Different models.

| Model | Pigments | Need of | | COMMENTS |
		Coefficients	Number Trials	
Quadratic	3	6	≥ 6	Not adapted to the colour
	4	10	≥ 10	
	5	14	≥ 14	
Special Cubic	3	7	≥ 10	Not adapted to extensive fields
	4	15	≥ 19	Rather precise for propositions
	5	29	≥ 35	
Cubic	3	10	≥ 12	Extensive fields
	4	20	≥ 24	Good precision DE<2
	5	35	≥ 45	
Quartic	3	15	≥ 19	High precision DE<1
	4	35	≥ 43	Equilibrium
	5	47	≥ 57	DL~Da~Db

Istanbul 15-19 May 2005

Quadratic Model.

$$L^* = f_L (a,b,c,d,ab,ac,ad,bc,bd,cd)$$

$$a^* = f_a (a,b,c,d,ab,ac,ad,bc,bd,cd)$$

$$b^* = f_b (a,b,c,d,ab,ac,ad,bc,bd,cd)$$

The function contain 10 terms, a minimum of 10 experiments will be needed.

Precision of calculation : 5 < DE < 7

Istanbul 15-19 May 2005

Special Cubic Model.

$$L^* = f_L (a,b,c,d,ab,ac,ad,bc,bd,cd,\mathbf{abc,abd,acd,bcd,abcd})$$

The function contains 15 terms, a minimum of 15 experiments will be needed

Experience shows 19 trials give better results.

Model giving good results when pigments are close in colour.

Istanbul 15-19 May 2005

Cubic Model.

$$L^* = f_L (a,b,c,d,ab,ac,ad,bc,bd,cd,abc,abd,acd,bcd,abcd$$
$$ab(a\text{-}b),ac(a\text{-}c),ad(a\text{-}d),bc(b\text{-}c),bd(b\text{-}d),cd(c\text{-}d))$$

The function contains 20 terms, a minimum of 20 experiments will be needed
Experience shows 24 trials give better results.

Model gives good results   DE < 2.

  Istanbul 15-19 May 2005

Quartic Model.

$$L^* = f_L (a,b,c,d,ab,ac,ad,bc,bd,cd,abc,abd,acd,bcd,abcd,$$
$$ab(a\text{-}b),ac(a\text{-}c),ad(a\text{-}d),bc(b\text{-}c),bd(b\text{-}d),cd(c\text{-}d),$$
$$a^2bc,a^2bd,a^2cd,b^2cd,ab^2c,ac^2d,bc^2d,abc^2,abd^2,$$
$$ab(a\text{-}b)^2,ac(a\text{-}c)^2,ad(a\text{-}d)^2,bc(b\text{-}c)^2,bd(b\text{-}d)^2,cd(c\text{-}d)^2)$$

The function contains 35 terms, a minimum of 35 experiments will be needed
Experience shows 43 trials give better results.

Model gives high precision   DE < 1.

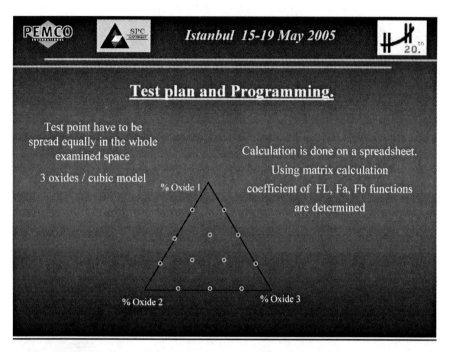

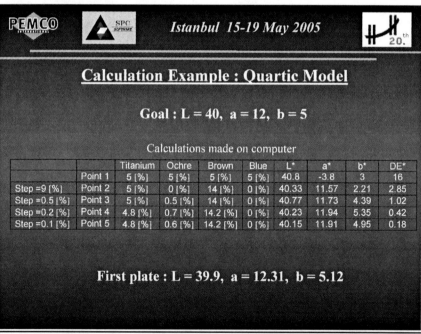

Istanbul 15-19 May 2005

Limitations and Advantages

- Limited colour space : each set of pigments defines a possible colour space, in which the target colour needs to be.

- Number of experiments : in order to achieve reasonable precision, the mixture plan requires numerous trials.

- Metamerism : calculation is made using CIE L*, a*, b* co-ordinates, the full spectral curve of colour is not taken into account.

- Accuracy of calculation can be very good as shown in previous example.

  Istanbul 15-19 May 2005

Colour matching & computer assistance.

A different way of predicting colour is using

the Kubelka - Munk theory

SPC has developed a program using the principles of this theory,

and able to predict colour

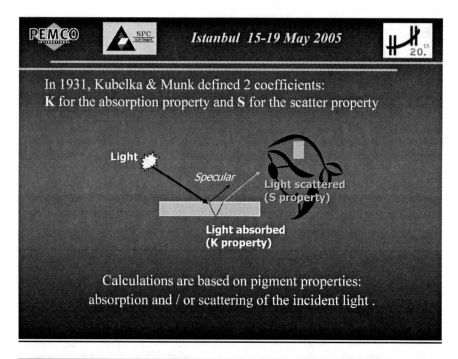

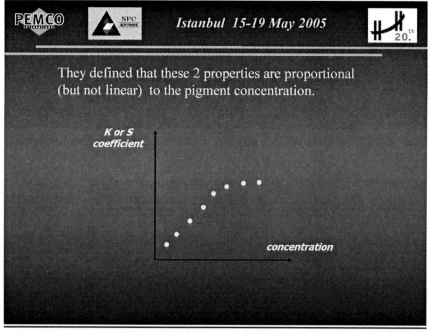

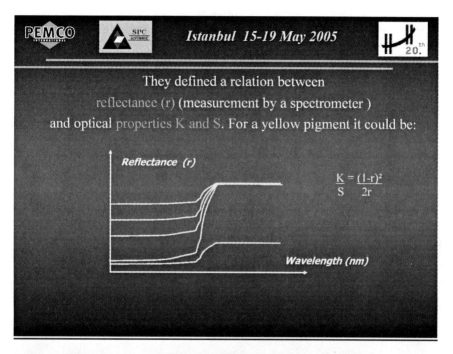

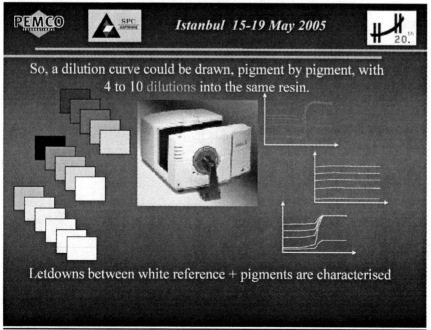

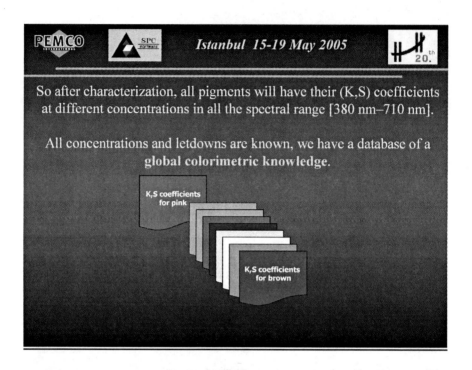

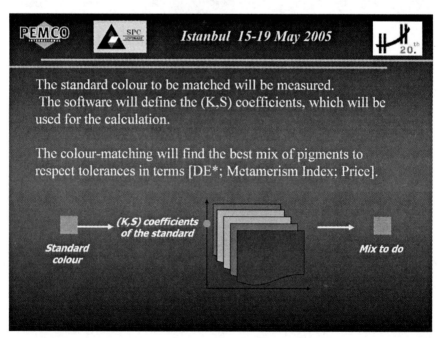

Istanbul 15-19 May 2005

The colour-matching could respect different constraints: [DE* Metamerism Index; Price], nature of the resin, pigment loading, opacity ratio ...

Let me show you an interactive demonstration in IsoMatch™.

  Istanbul 15-19 May 2005

Colour matching using IsoMatch™

Using a set of 22 silk screen pastes a data base has been produced.

Letdowns of white SD01
with all SD pastes
have been prepared

K and S of all products have been
generated

Colour calculation can then be run

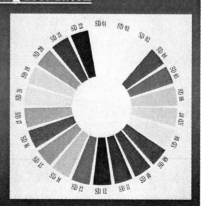

FERRO NEW PRODUCTS FOR THE HOME OF TOMORROW

Charles Baldwin, Brad Devine, Dave Fedak, and William Warner
Ferro Corporation

Abstract
As mature products, porcelain enamel and, in general, major appliances require technological advances for added sales and profitability. Three new porcelain enamel products and their most suitable uses are reviewed.

Product Life Cycle
Figure 1 shows a plot of product life cycle versus sales and profits.[1] There are four stages. First, during the introduction phase, there is typically a product development phase where profits may be negative as research and development costs are recovered by initial sales. The product should be carefully monitored and can be withdrawn if growth is not reached in an acceptable timeframe. Second, if the product sells successfully, there is a stage of rapid growth as orders quickly increase. Output rapidly surges, and the price can be set high if there is little competition. Portable digital music players such as the iPod are at this stage. Maturity, the third stage, is the most common and the one at which both porcelain enamels and major appliances are. Marketing and finance are the most common activities. Competition can be fierce, and research and development activities are focused on incremental improvements or increasing efficiency. Finally, in the decline stage, perhaps from technological obsolescence, the market begins to shrink as do sales and profits. The choice may be made to discontinue the product. VCRs and typewriters would be in the declining stage.

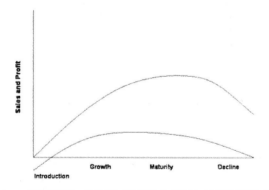

Figure 1. Product life cycle[2]

While porcelain enamel usage remains strong on major appliances and in related industries (plumbing ware, barbeque, cookware, and so forth), a look at major appliance milestones listed in Table 1 suggests a need for innovation.

Year	Milestone
Mid 1850s	First gas range
1893	First powered dishwasher
1924	First white enameled range
1963	First self-cleaning ranges

Table 1. Appliance milestones[3]

Furthermore, the percentage of surface area of a range that is enameled has fallen quite a bit since the first white enameled oven came off the line in 1924. A significant portion is powder coated. Alternatively, stainless steel is used. Refrigerators, dishwashers, and clothes dryers are rarely enameled today.

Two new features engineered into porcelain enamel are enhanced cleanability and metallic appearances. For cleanability, wipe-clean release was created for temperatures below 600°F (316°C), and alternatives to self-clean (and continuous clean) enamels were developed for the range past 600°F (316°C). For modernizing the appearance, metallic enamels have been formulated that capture the appearance of stainless steel while retaining the mechanical and thermal durability of the porcelain.

Wipe-Clean Release Enamel
RealEase™ combines the cleanability of the organic non-sticks with the durability of vitreous enamel. It offers the scratch resistance of enamel and the cleanability of PTFE. It is a patented technology that stands to bridge the gap between the organic non-sticks and low-temperature porcelain enamel.

In terms of features and benefits, the improved cleanability saves consumers time and effort. Because the coating is difficult to scratch and heat resistant, it retains its original appearance longer than PTFE. Its ceramic nature also suggests environmental friendliness.

Cleanability was evaluated using the Easy-to-Clean (ETC) test, developed by Ferro-France, by cooking egg beaters, ketchup, salted whole milk, lemon juice, and gravy separately onto the coating at 450°F (232°C) for one hour and assessing the force needed to remove each soil. A score of 5 was received if the soil was easily removed, and 1 if it could not be removed for a maximum possible score of 25. The results are shown in Figure 2. The cleanability of traditional enamel exceeded that of stainless. In turn, RealEase™ matched PTFE and out-performed enamel.

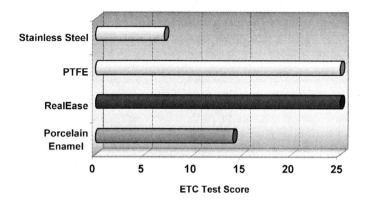

Figure 2. Cleanability test results

The hardness was measured using ASTM D 3363-00 "Standard Test Method for Film Hardness by Pencil Test". Different calibrated hardness pencils are worked through from the hardest to the softest and the one that will not rupture or gouge the coating is reported. The range, from softest to hardest, is:

6B - 5B - 4B - 3B - 2B - B - HB - F - H - 2H - 3H - 4H - 5H - 6H – 7H – 8H – 9H

Figure 3 shows the results of pencil hardness testing conducted on a PTFE coating and a RealEase™ coating at room temperature and in a pan left on a high flame on a gas range for 30 minutes. Note that the PTFE coating softened considerably. Therefore, during cooking, any metal utensils would be more likely to severely damage the coating.

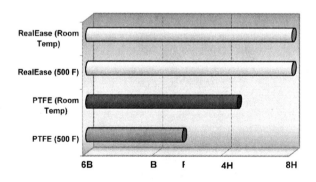

Figure 3. Pencil hardness test results

With superior cleanability, scratch resistance, and heat resistance, potential applications for a wipe-clean porcelain enamel usable up to 500 to 600°F (260 to 316°C) can be considered. For domestic uses, these include cookware and bakeware (aluminum, aluminized steel, stainless steel, or ceramic), small appliances, toaster and microwave ovens, simmer plates, and outdoor/backyard grills/griddles. Because RealEase[TM] is certified as safe for use in restaurant kitchens, it is also suitable for commercial kitchenware, cookware, bakeware, and appliances.

Steam-Clean Enamel
Oven cleaning technology consists of three types: (1) self-cleaning pyrolytic ground coat, (2) non-self-cleaning ground coat, (3) catalytic continuous clean enamels.

For the self-cleaning enamel, food residue is reduced to ash by exposure to temperatures between about 900 and 1000°F (482 and 538°C). This is the leader in the North American market because of minimal consumer interface with the oven to remove the ash after the cleaning cycle, but there are several concerns with it. First, high heat is required, necessitating extra insulation and safety interlocks. Second, there are concerns about the possible release of toxic fumes.[4] Third, the insulation also makes the addition of advanced electronics (e.g., to integrate the range into a home network) more difficult. Finally, to meet the requirements of surviving multiple clean cycles, the enamel generally contains hard, chemically-resistant frits that, without the high-temperature exposure, have poor release properties on their own.

Non-self-clean enamel requires significant effort by the consumer and/or harsh alkaline saponifying cleansers (e.g., Easy-Off®) that have a pH of approximately 14[5].

Catalytic continuous clean enamels contained specially formulated glasses, which had high levels of metal oxides. These fired out with a porous microstructure enabled the reduction to ash at normal cooking temperatures. These have largely fallen out of use in North America.

AquaRealEase[TM] is a porcelain enamel with a patented formulation that allows baked-on food residues to be released with exposure to moisture (either as water or steam). As such, it addresses many of the problems with the self-cleaning enamel. It has the mechanical durability and thermal resistance of traditional enamel, can be applied in a single fire to steel, and fires out between 170 and 1570°F (799 and 854°C).

Figure 4. (a) Plates with baked-on foodstuffs (a) before and (b) after exposure to warm soapy water for 15 minutes

Figure 4 shows the result of a cleanability test. A self-cleaning ground coat panel is on the left and an AquaRealEase™-coated panel on the right. Clockwise from the upper left, the soils were a 50:50 egg:oil blend, French vanilla cake mix, cherry preserves, and Hollandaise sauce. These were baked on at 450°F (232°C) for one hour in a gas oven. The panels were then soaked in warm soapy water for 15 minutes. The soils were all released from the AquaRealEase™-enameled plate. On the self-cleaning enamel, the Cherry preserves could not be removed even with vigorous scrubbing and abrasive cleaners.

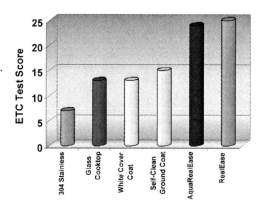

Figure 5. Cleanability results for appliance finishes

To obtain quantitative data, the easy-to-clean test was run on alloy 304 stainless steel, a plate of Ceran glasstop material, a typical titanium-opacified cover coat, self-cleaning ground coat, AquaRealEase™ and RealEase™. Results are shown in Figure 5. After the moisture exposure, the release properties of the AquaRealEase™ equaled RealEase™ and significantly exceeded those of pyrolytic ground coat.

High-Temperature Metallic Enamel
Previously, a comparison of the mechanical, chemical, and thermal resistance of porcelain enamel and stainless steel was made that concluded that while the appearance of stainless steel is very popular, porcelain enamel is more suitable for an appliance finish. While stainless steel has a pencil hardness of 5H and a Rockwell Hardness of 88 HRB (roughly equivalent to a Mohs hardness of 4), the pencil hardness of sheet steel porcelain is so hard it is off the scale and it has a Mohs hardness of 5 – 6. Furthermore, stainless steel plates significantly yellowed and darkened after exposure to 750°F (399°C) while the enamel did not show any signs of discoloration at all.[6]

Test	White Cover Coat	Stainless-Style
Taber Abrasion		
m_i (g)	74.42	75.54
m_f (g)	74.41	75.53
Taber Abrasion	10 mg weight loss	10 mg weight loss
Boiling Acid	4.55 mg/in^2	1.83 mg/in^2
Boiling Alkali	18.65 mg/in^2	0.86 mg/in^2
Spot Acid Resistance	AA	AA
Thermal Shock	No chipping	No chipping

Table 2. Comparative test results between a white cover coat and the stainless-style cover coat

Table 2 shows results of comparative testing between a conventional titanium-opacified cover coat and the stainless-style cover coat. First, the materials were tested according to ASTM D 4060-95 "Standard Test Method for Abrasion Resistance of Organic Coatings by the Taber Abraser" and then ASTM C 282-54 "Resistance of Porcelain Enameled Utensils to Boiling Acid" for 6 hours with both 6% citric acid aqueous solution and 5% tetrasodium pyrophosphate solution. Third, the fired enamels were evaluated per ASTM C 282-99 "Standard Test Method for Acid Resistance of Porcelain Enamels (Citric Acid Spot Test)". Lastly, the panels were exposed to three heating cycles at 600°F (316°C) for 30 minutes followed by immersion into room temperature water to test thermal shock resistance. With such properties, the metallic enamels can be used for a wide range of applications including surfaces currently coated with enamels firing in the 1500°F range such as cooktops, ranges, sanitary ware, refrigerator exteriors, steel or cast iron cookware and bake ware, car mufflers, and barbecue grills.

Summary

Three new enamel technologies were reviewed: non-stick enamel RealEase™, water-cleaning enamel AquaRealEase™, and stainless-style metallic cover coat. First, with a firing temperature and cleanability similar to PTFE coatings but superior mechanical and thermal resistance, RealEase™ would be most suitable for use on pots, pans, bakeware, griddles, and small appliances. Because of the difficult to scratch the material and a lower tendency to discolor on exposure to heat, the original appearance would be maintained longer than PTFE. Second, AquaRealEase™ offers an oven-cleaning mechanism free from the necessity of high heat and fumes. Third, metallic enamels would permit enamel to have new appearances and be more suitable for high-end products. These three new products create the potential to move to the growth portion of the product life cycle curve.

References

[1] "Product Life Cycle Management"
Wikipedia, The Free Encyclopedia
http://en.wikipedia.org/wiki/Product_life_cycle_management
(June 12, 2006).

[2] "Marketing – Products – Product Life Cycle"
http://www.tutor2u.net/business/marketing/products_lifecycle.asp
(June 12, 2006)

[3] "Appliance Milestones"
http://www.aham.org/consumer/ht/action/GetDocumentAction/id/1408
(June 25, 2006)

[4] "Tips from eHow users on How to Clean an Oven"
http://www.ehow.com/tips_2017.html
(July 31, 2006)

[5] RECKITT & COLMAN INC. 1993. *Low temperature non-caustic oven cleaning composition.* United States Patent 5,380,454. 1993-07-09.

[6] Dave Fedak and Charles Baldwin, "A Comparison of Enameled and Stainless Steel Surfaces," *Proceedings of the 67th Porcelain Enamel Institute Technical Forum*, 45 – 53 (2005).

Wet Porcelain Enamels and Processing

CONVEYOR ROTATORS

Richard Dooley
AP Conveyor
Toledo, Ohio

Abstract
This paper reviews current designs in conveyor rotators to help increase line utilization density. Rotators are devices on the parts hangers that allow for the pieces suspended to be rotated various angles to facilitate close packing as with power-and-free conveying systems, increase exposure during material application to substrates, improved loading and unloading access, and movement up and down inclines. Some of the rotators are static assemblies and some are actuated mechanically.

Introduction
For a variety of reasons, it is sometimes desirable to rotate parts or hangers (racks) about a vertical axis during their passage through a finishing system. Some of these are:

• Loading and unloading access from a single side of the conveyor.
• Preventing rack-to-rack contact in the accumulation zones for high speed power-and-free systems.
• Facilitating close packing in ovens for power-and-free systems.
• Spraying both sides of a rack from a single side of the booth (1 80° rotation in booth).
• Spraying one side of flat panels mounted on a square ware package rack (90° rotation in booth).
• Improving spray access to racks mounted perpendicular to line (45° rotation in booth).
• Continuous rotation of radially symmetrical parts in corona zone of booth.
• Increasing clearance between parts on conveyor inclines and declines.

The common terms for devices that accomplish this are Rotators and Indexers, sometimes used interchangeably. "Rotator" tends to apply more to continuous or intermittent motion in one direction, while an "Indexer" may oscillate back and forth between two detented positions. Either one may be manually or mechanically actuated, with carrying capacity ranging into thousands of pounds.

Skate wheels, sprockets, star wheels, and torpedoes (Fig. 1 through 4) are the classic rotator designs. All have been in wide industrial use since World War II.

Figure 1 – Skate Wheel Rotator (Courtesy of Mighty Hook, Inc.)

Figure 2 – Sprocket Rotator with actuator track

Figure 3 – Star Wheel Rotator (Courtesy of Richards-Wilcox)

Figure 4 – Torpedo Rotator. Horizontal tracks engage torpedoes at different altitudes to provide 90° incremented rotation

Supplementing these have been custom designs with varying degrees of suitability for paint line use (Fig. 5 & 6).

 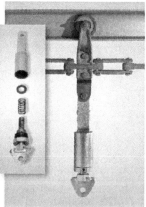 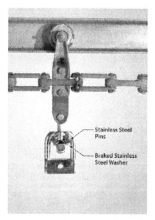

Figure 5 – Star Wheel Rotator with internal bearing and detent

Figure 6 – Spring Loaded Manual 180° Rotator, with internal mechanism

Figure 7 – Manual 180° Rotator. This unit replaced the Figure 6 unit in production use. Note open construction for drainage.

Rotators with enclosed mechanisms and carbon steel wearing surfaces can trap washer solutions which will corrode and bind up the mechanisms, as well as spitting residue onto the parts below as they pass through the ovens. Carbon steel components will also soften gradually at pyrolytic stripping oven temperatures. The limitations of these existing designs are being bypassed by a new generation of rotators. Units providing detents of variable stiffness, positive latching, controlled powered rotation at high line speeds, and 90° to 60° sorting, are being used in production spray finishing applications.

Detents

Detents used in this new generation of rotator designs have a braked Mil-Spec stainless steel washer and two stainless steel pins as their common elements *(Fig. 7)*. By varying the angle at which the washer is bent down from horizontal, the stiffness of the detent may be controlled. For manual actuation, 7½° is most commonly used, although a reduction to 5° was necessary in one application involving wheels for large aircraft *(Fig. 8)*. At 60°, the rotator will self-lock, requiring the rack to be lifted out of the detent position before turning. If positive locking of a softer detent is required, a separate gravity latch, disengaged by a glide bar flanking the conveyor, is used to keep the pins from climbing the flanks of the washer *(Fig. 9)*. Lateral notches are milled into the pin lower surfaces for a secondary detented position at 45° or 90°. At times it is advantageous to build the rotator into the rack itself rather than having it as a separate element *(Fig. 10)*.

Figure 8 – Ninety Degree Detented Manual Rotator with 5° braked washer for use with heavy workpieces

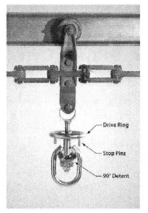

Figure 9 – Manual 180° Rotator with gravity latch. Glide bar flanking conveyor lifts latch at unload-load areas

Figure 10 – T-Bar Rack with integral rotator

Figure 11 – Powered 90° Rotator

Powered Rotation

When automatic rotation or indexing is required at a specific location on the conveyor, a ramp assembly is used in conjunction with a drive ring on the rotator. This ramp lifts under one side of the ring to raise the detent out of engagement, then "drag" the ring around by friction on the bottom and sometimes on the outside diameter, into the alternate position. The rotation is stopped by a vertical pin protruding from the ring, which contacts the inboard side of the lift track, or a separate stop track in some cases *(Fig. 11)*. The tracks may be designed to slope as gradually as required, so actuation is very smooth even at the highest line speeds of 65 + ft. per minute. When the rotation is stopped, the lift track slopes back down to reengage the detent. Powered rotations of 45° to 1 80° are accomplished by these mechanisms. An additional advantage, particularly for robot spraying, is that a rack coming through the actuator tracks in the wrong position will not be moved from that position, so racks down-line of the actuator track will all be correctly oriented.

For sprocket rotators, a positive drive that will rotate parts rapidly in the booth corona zones involves the use of a chain and sprocket drive system *(Fig. 12)*. This provides the averaging effect of multiple passes on cylindrical work pieces such as water heater jackets, compressed gas storage cylinders, and fire extinguisher bodies.

Figure 12 –
Powered
Continuous
Rotator
(Courtesy of
Protectaire
Finishing)

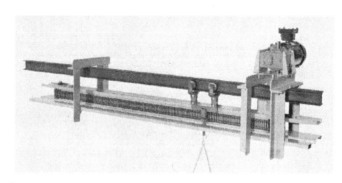

Forty five degree rotation for spray or racking access may also be accomplished by a variation of the Angle Pivot™ mechanism. In this, the rack is lifted and rotated by a glide bar, which moves an actuator rod to its alternate position *(Fig. 13 & 14)*. Detent in the normal hanging position is provided by gravity.

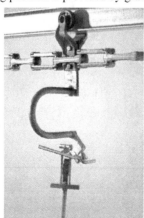 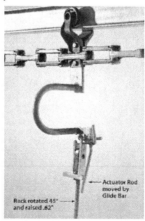

Figure 13 – Angle Pivot™ 45° Indexer in
running position

Figure 14 – Angle Pivot™ 45° Indexer in
actuated position

Use With Load Bars

Combining rotators with Angle Pivot™ continuous load bars on the same system provides the line with the advantages of each. With the in-plane thrust transmitting capability of the bars, the rotator and rack do not "flinch" when the powered drive mechanism engages. Racks that are rotated manually for loading and unloading are returned to the in-plane position by a set of tracks which contact a flag at the down-line end of the unload/load station *(Fig. 15)*.

Figure 15 – Angle Pivot™ C-Hook with manually activated ± 110° rotator, installed on continuous load bar

For 80° rotation in the booths, a drive ring and 60° detent washers are used *(Fig. 16)*. Sprocket rotators for use in the booth corona zones have the tooth contact point in axial alignment with the Angle Pivot™ bearing. This also reduces the recoil when the power driving mechanism engages *(Fig. 17)*.

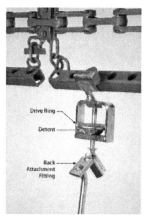

Figure 16 – Powered 180° Rotator on continuous load bar

Figure 17 – Sprocket Rotator on continuous load bar

A variation which combines sprockets with a 80° detent is used to index aluminum extrusions at the halfway point in a Ransburg Omega booth *(Fig. 18)*. A manual rotator with a hand-operated latch has been used for single-side loading and unloading of long parts such as moldings and trailer hitches. Here it is desirable for the parts to be processed through pretreatment tipped 3° – 5° backwards, so the process solutions will drain back into the proper stage. A notched washer at the rotation bearing prevents the latch from engaging in the reversed position *(Fig. 19)*.

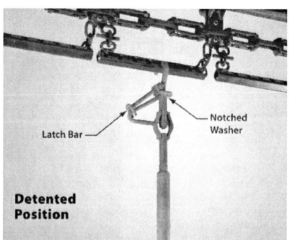

Latch Bar —

— Notched
Washer

**Detented
Position**

Figure 18 – Detented 180° Sprocket-Powered Rotators on continuous load bar. Pitch between rotators is 2". Used on vertical extrusion painting lines.

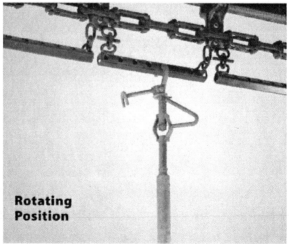

**Rotating
Position**

Figure 19 – Manual Latching 360° Rotator for single-side loading and unloading of long parts

The fanciest rotation scheme attempted to date is shown in *Figures 20, 21, and 22*. It involves a latched 80° detented rotator with manual and powered actuation, plus an Angle Pivot™ mechanism to permit negotiating 45° inclines and declines. The pivot angle used was so steep that it was necessary to immobilize the Angle Pivot™ bearing during rotation, to prevent the two from interfering with one another. The powered rotation drive was taken by friction from the conveyor chain, and imparted to the drive ring O.D. by the back surface of two Poly-groove belts, driven at different speeds. As with all of our actuation tracks, a quick-release mechanism will deactivate the drive, and permit a stuck rotator to be quickly cleared.

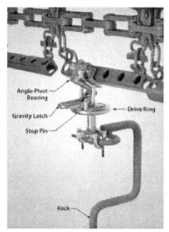

Figure 20 – Latching Powered and Manual 180° Rotator with Angle Pivot™ bearing

Figure 21 – Actuator Track Assembly for Figure 20 rotator, friction driven from conveyor chain

Figure 22 – Actuator Track Assembly in retracted position

Conclusions

Rotators with many additional features are available to complement the skate wheels, star wheels, sprockets, and torpedoes of yore. Initially built integrally with paint racks, they are now offered as a stand-alone product to provide cost-effective versatility to paint and porcelain spraying systems.

AUTOMATED ROBOTIC COATINGS AND FINISHING SYSTEMS

Jerry L. Perez
FANUC Robotics America, Toledo, Ohio

Abstract
Automated robotic finishing systems of today can implemented for many different types of coatings applications. In the past 5 years, major developments had been made to allow for robotic finishing systems to land themselves in many new markets. There have been large advances in part identification and robotic programming techniques and technologies. Manufacturers will find that although in previous years, robotic application of wet spray porcelain enamel may not have been the most ideal application for their system, today, they will find that robotic applications could now be the right solution for the application needs.

Where We Have Been Before
In the past, manufacturers found large limitations when they attempted to implement robotic applications for their finishing systems. In the past, if the plant ran a low volume job shop with a lot size of one, they may spend more time creating robot programs versus painting their parts. Also in the past, automated fluid delivery systems could not maintain the consistent flow and could not respond to required pressure step changes. With breakthroughs in automatic part identification and advances in material delivery technology, there is now a larger potential market for the robotic application of porcelain enamel. With new developments in robot manufacturing techniques and robot technology, the costs for implementing these systems have lowered.

New Technologies Assisting the Use of Robots

Automatic Part ID
Vision systems and laser scanning systems of today, coupled with faster processors, have made the part identification process easier. In the past, some plants could not implement robots simply because they could not reliably input all of the different part styles they had to paint – they just had too "much" variation. Today 3D scanning technologies eliminate the need for manual part ID input.

Automatic Path Generation
Using data collected from the Automatic Part ID system, robot paths can now be created automatically with third party software in Europe and in the United States. Depending on part complexity, the paths generated could accomplish 100% coverage without any type of manual touch up. This new technology eliminates the hours needed for creating robot programs manually when a plant has a large variety of parts to coat.

Non-Contact Material Delivery System
Most fluid systems of today use pressure pots or dual diaphragm pumps to delivery wet porcelain glaze or enamel to the spray gun. There is limited variability in flow with this type of setup and typically, the operator is limited to using one setting for all part styles. Using pressure pots or

diaphragm pumps required complex programming techniques from slowing down and speeding up robot tip speeds and increasing and decreasing target distances. Today, there is new technology that provides non-contact high precision positive displacement fluid delivery and fluid control. This system has an approximate step response time of 1ms which allow for the user to apply the material in one area at one specified fluid flow and then on the next pass apply a completely different fluid flow rate. The technology is based on servo-motor driven pumps which are never in contact with the material itself. With the high precision and consistent flow rates achieved through the servo motor controller pumps, and the durable nature of a non contact system, this system now allows users to program at one robot speed and target distance. This new fluid delivery and fluid control system enhances a robotic system's finish quality and coverage consistency, and reduces programming time.

Developments at the Robot Manufacturer

Lean Manufacturing Techniques
Better robot manufacturing techniques have decreased costs for robots, opening the doors to many new users in many new markets. FANUC Robotics manufactures their robots with their own robots – robots building robots. By practicing today's lean manufacturing techniques, FANUC has continued to improve on their products costs and are able to provide products at lower costs to many new potential clients.

Robot Controller Development
The robot controller has become more robust and has high capability today. The controller can control end of arm tooling, spray equipment, and fluid delivery equipment through its own software and processors – no third party software, controls, and control panels required. New technologies, such as vision technologies are also integrated in the robot controller, again making it easier and cheaper to use. Multiple robots can be run off of one controller now, decreasing costs up to 20% per systems with multiple robots. There is also an increase in the accuracy of programs taught in a slower teach mode. Typically, the robot is taught a program at slower speeds and run during production at higher "production speeds." In the past, there could be considerable "touch-up" work required to adjust the programs to avoid "rounding corners" when running at higher speeds. Today, a program taught at 400 mm/s will run the same path at 1500 mm/s without touching up and adjusting the paths. All of these advances and new capabilities make it easier to implement robots and more cost effective.

New Potential Robotic Users

Reasons to Use Robots
The typical reasons a plant should consider to use robots are as follows: their material costs continue to rise, their labor costs continue to rise, their customers are demanding high quality and higher throughput, they have a high turn over rate in the finishing department, there is over 10% scrap and rework currently, and they run multiple shifts. If you find yourself in more than one of these categories, you should consider robotics.

Benefits from Using Robots

The typical benefits a plant would realize from using robots are as follows: lower material usage, lower labor costs, higher quality, healthier work environment, increased throughputs, and lower scrap and rework amounts.

Accurate Trigger Times = Lower Material Costs and Higher Quality.

Manual applications will vary +/- 10% from the optimal amount of material per part whereas robotic application will vary 1.5% above and below that number.

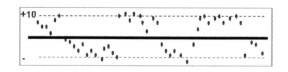

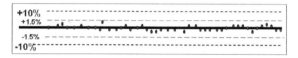

Accurate Trigger Times = Lower Material Costs and Higher Quality.

Manual applications will vary +/- 10% from the optimal amount of material per part whereas robotic application will vary 1.5% above and below that number.

Typical Example savings by

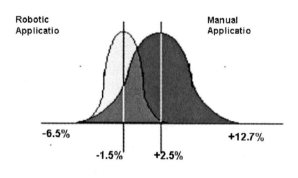

Robotics to Save Your Factory

Investments up Front Pay off in the Future

Altough the paradigm of manufacturing plants today is it is cheaper to manufacture overseas, robotics coupled with lean manufacturing techniques, can save your factory and actually beat your overseas competition in the long run.

North America

Offshore

- Lean Manufacturing
- Robots & Automation
- Quality Programs
- Just in time Inventory
- Capitalize on culture, laws & standards

STATE OF THE ART OF THE ENAMELLING OF CAST IRON GRATES

Harry Florijn
Lovink-Technocast b.v.
The Netherlands

Abstract
Founded as a foundry in 1911, Lovink started producing grates around 1950. The enamel shop started in 1979. The enamelling of grates started in 1980 by wet spraying. More then 25 years of experience brought us in the high end of the cooking appliance markets around the world. The quality of these enamelled castings is the benchmark in this market sector.

Over the years, many different enamel application methods have been tried for improvement of quality or reduction of production costs. Of all application methods, the EPE application is "State of the Art".

Markets developments
In the appliance market for gas ranges and cook tops we have seen some development trends over the last 15 years. One of the most important has been a strong trend towards stainless steel in combination with the color matt black for grates. This made it for the competition easier to close a part of the quality gap in comparison with the glossy colors black, taupe and grey when EPE was compared with other application methods. Other trends are the increase of market share for ceramic, inductive cooking, and cast iron grates. Gas on glass with more design destinction. And of great importance now, as can be observed in the increase in lectures about these topis, also strong trends towards metallic colors and more smooth and better cleanable matt black surfaces. It did cost Lovink a lot of effort to widen up the quality gap again over smoothness of finish, metalic colors and cleanibility over the competition. Gas will play an important role now and in future.

The market for grates consists of a few classes with different base materials and with/without the enamel coating. From wired stainless steel grids (no enamel) to pressed enamelled steel grids, wired and bar steel enamelled grids and single, double and more exotic designed enamelled cast iron grates. There is a quality and price increase in the same order. Amongst grate quality features, color and gloss, heat resistance, thermal shock, chemical resistance, cleanability, and stability of the shape are the most important. Looking at all combined quality features for the coating, enamel is still the only coating material that is really suitable here.

Application methods
Wet spraying, dipping, flow coating were the first wet methods applied on grates. Later wet electrostatic methods were tried. Also at approx the same time the EPE process became available. And a little later the dry methods, elecrostatic and fluidised bed. Each method tells its own story on grates.

Dipping (with movements) gave visual drainlines, uneven application and it as an application method not suitable for parts where visual pleasing surfaces are requested. Appropiate set up of enamel in combination with thickness controll is in pratice not well achievable. This also applies for flow coating processes. Electrostatic wet spraying can be automised and has helped to reduce labour costs and enamel costs, however the restrictions of the electrical fields are wellknown. The EPE process is however a good method for grates.

From the dry methods, powder spraying is the best known. It can be automised easily and it shows a high reclaim rate. However the traditional disadvantages of the dry powder process work out particular strong in the case of grates. Less direction controlled enamel stream on planes not perpendicular on the enamel flow gives uneven apllication thicknesses. Handling problems, limitation in color, very biscuit sensitive, adherence problems. In 1C/1F situation, problems in the matt finish for consistent quality. Cage of Faraday effects. Hardly impossible to bring about the requiered thermal (shock) resistance. Fluidised bed coating with heated castings gave problems with the heat capacity in the thinner sections of grates. And it also proved to be very difficult to automate.

A process that is automated, has high enamel yields and delivers technical and visual good enamel finishes is the EPE process.

EPE as a process
Within the PEI, reports over the EPE enamelling of grates date from 1992 (an EPE update in general). In 1995/1996 (Vermont/Whirlpool and Vermont/Ferro). In 1998 (Eisenmann) and in 2003/2004 (Roesch/Ferro).Emphasis was given all the times on cast iron grates. However in the market no striking successes were reported in the US.

A good definition of the electrophoresis process was given: The movement of suspendid particles through a fluid under the action of an electromotive force applied to electrodes in contact with the suspension. In practice the following happens: The workpiece (grate) is an electrode, that act as an anode (+ potential), thus attracting the negative loaded particles.The other side collects the water (ions) and is the electrode/kathode (-potential). Water is squeezed out the enamel layer to a content of 20%, with a very distinctive separation layer. The part can be handled by man to transfer it to the furnace. Taking enamel out (with the product), also means taking water out. This is done at the kathode side, with a dialysis cell (with a semi-permeable wall). Also oxygen and hydrogen gases develop at anode and kathode side.

In production there are a numbre of consequent process steps, that differ from application to application. But allways a conductive layer is created.

From the enamel side a few controll steps are important:

-Enamel should have a sufficient low conductivity and high specific weight, so in production process adjustments are possible
-The amount of coloring oxides or opacifiers should be as low as possible, get colors from the fritts

-Have good (electrical) stability from the applied enamels as a pre-condition (limited soluable in water). The process starts and ends with a full bath, so storage (time, temperature) should have very limited effects on the electrical properties of the enamel sludge.

-Balanced particle size distribution

-Addition of sodium aluminate gives additional controll of the uneven layer thickness distribution. It is a question of deleting field lines or better spread them equally. The deposition of the builded Al(OH)3 builds up quickly resistance in the enamel layer.

Other important controll items:

-Drain controll (by ultrafiltration/microfiltration) gives a way of succes to further controll the electric properties of the enamel and a way to increase the efficiency of the enamel usage (70-90%)

-Strict conditions on the castings and casting surfaces: holes canot be filled, roughness effects, problems with recepees with multiple enamel sorts.

-Good agitation controll

-Temperature controll

-Level controll

Conclusions

There are some very important advantages in the EPE application methods that are particular of interest when enamelling cast iron grates:

Complicated shapes are no problems. Basically, the enamel is allready there, where it should be, close to the casting surface. The edge coverage is excellent. Also in case of matt enamels a smooth surface finish can be obtained. The parts can be covered all around, giving a thinner more uniform coating of very well controlled thickness. This is a great help in the overall corrosion resistant protection and in creating dish washer proof products.

However there are also a few complicating factors:

-The process controll is very complex, and should be always present

-High investment costs

-Compared to other application technologies EPE is much more knowledge based, skilled people should be hired

-Changes in enamels urge research on many fields

-Development times for new enamels are very long, so working ahead of marketing demands is a big issue

-There are some physical limitations that are ever present: hanging during firing (deformation of the castings), speed of transport of the particles and products (cycle time), oxygen development at the anode position, cage of faraday effects.

After years of trials and steady progress, the EPE process continues to be "State of the art" for the enamelling of cast iron grates.

MEETING FUTURE REQUIREMENTS FOR ENAMELED CAST IRON GRATES

Ron Simons
Lovink-Technocast b.v.
The Netherlands

Abstract
Over the last decades, the quantity of enameled cast irons grates has been growing significant in the appliance industry. Besides this growth, a trend towards shorter cooking times is noticed, leading to high power burners. Due to this fact, the requirements have been increasing slightly but also continuously.

In this lecture an overview is given for the current requirements for enameled cast iron grates, and the expectations into this area from Lovink for the near future.

The requirements
The requirements for supplying cast iron enameled grates on the current global market is not limited to quality aspects only, although this is still be thought of being of main importance. Other aspects which are being meant are logistical and development related.

Concentrating on the quality aspects, next parameters are found to be important :

a) Visual aspects :
   Improving the visual appearance towards visual "perfect" grates is a goal. More and more small failures are no longer tolerated, although the functionality of the grates is not violated.
b) Flatness :
   The currently achieved flatness of the grate itself as well as flatness over the fingers are expected to be sufficient for now as well as in the near future.
c) Acid resistance :
   The acid resistance test (using a citric acid solution) gives a good first qualification for resistance against food over life time usage. It is expected that the currently being used method will be maintained. Life time testing at customer sites differs very much, and will therefore always be additional required.
d) Thermal Shock resistance :
   Improvement will be required as faster heating-up times are being asked.
e) Impact resistance :
   The current achieved level is sufficient. More knowledge for the real adhesion and its basic principles is a matter of investigation.
f) Color/gloss:
   Due to the increasing requirements, the currently accepted color and gloss ranges are going to be narrowed. In particular matt enamel surfaces, for which the human eye is more sensitive for deviations, is a point of investigation at Lovink.
g) Dishwasher resistance :

The dishwasher resistance of cast iron enameled grates is an item which is already important, but will be even more important in the future. Although several tests do exist, there is not 1 general accepted dishwasher testing method.
In order to be able to judge and compare all kind of different enamel systems, Lovink has a testing method under construction.

h) Thermal resistance :
For similar reasons as for the thermal shock resistance, the resistance against elevated temperatures will have to increase too. Temperatures above 500 °C will be no exception for future designs.

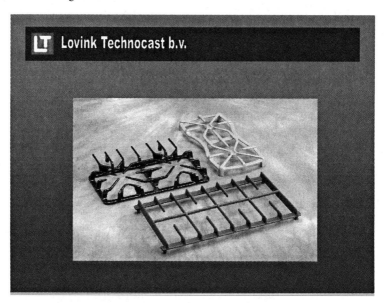

Other developments

Beside the above mentioned quality related items, Lovink coordinates developments of new enamel systems, leading to new looks, which are thought to become more and more industrial trend.

It is predicted that visual appearances, like metal look, cast iron look and paint look etc., and still having the well known advantages of the properties of enamel, must become available within the next decade. Therefore a significant amount of investment is being done by Lovink into this area.

MAINTAINING A RELIABLE WET ENAMEL SYSTEM

Dennis Opp
Whirlpool [Maytag]
Cleveland, Tennessee

Abstract
A successful enameling operating depends on specific process parameters and their maintenance. This paper illustrates the operation of the use of pre-milled enamel (ready-to-use) at Whirlpool's Cleveland, Tennessee plant. Attention is focused on maintaining the proper rheological characteristics of the enamel slurry through regular measurements of the slump, specific gravity, and pick-up of the enamel. Additionally, control of the metal cleaning operation is maintained as well as the dryer for the cleaned parts. Finally, the process evaluates the coating thickness after flow-coating by wet film gage. The result of good process control is consistently high quality throughput.

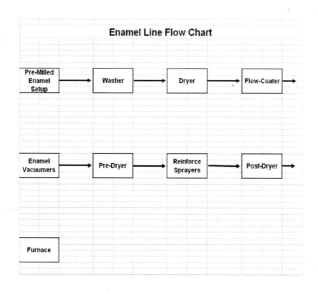

Enamel Line Flow Chart

Pre-Mill Enamel Set-Up

- Add Water and Stir.
- Age 24 Hours before using.
- Measure Slump.
- Check Gravity. (Adjust with water or enamel to specs).
- Check Pickup. (Adjust with Sod. Tripolyphosphate or Pot. Nitrite).
- Filter Enamel through 30 Mesh Screen.

Loading Enamel

Measuring the amount of Slump

PLANT 3 ENAMEL CHECKS					
DIP TANK					
Date	Time	Material	Gravity	Pick-up	Filled from Tank #
REINFORCE TANKS					
Date	Time	Material	Gravity	Slump	Spray Tank #
REWORK TANK					
Date	Time	Material	Gravity	Pick-up	Slump
Comments:					
					Form 0083

Ensure Optimum Washer Performance:

- Maintain a detailed cleaning schedule.
- Proper nozzle alignment. (Spray patterns, pressure).
- Daily checks to ensure concentration, soil load, water pressure and temperature.
- Periodically check parts for cleanliness after they exit washer.
- Acid clean washer as needed, (generally every four months).
- Side note: We use an Alkaline based liquid Cleaner and a ether based solvent, (only when we run into tough cleaning issues).

OVEN WASHER LINE DAILY CONTROL SHEET

Date:

Time	Prewash	Stage #1	Stage #2					Stage #3					Stage #4	Stage #5	Stage #6	Stage #7
	Clean Daily	Clean Daily	2.5-3.0 oz/gal	Cleaner Added	Press 8-10psi	Temp 140-180F	Soil Load	2.5-3.0 oz/gal	Cleaner Added	Press 6-8 psi	Temp 140-180F	Soil Load	Temp / Press 4-8psi	Temp / Press 10-20psi	Temp / Press 10-20psi	F. W. Halo 10-20psi
1st Shift													/	/	/	
													/	/	/	
													/	/	/	
													/	/	/	
													/	/	/	
2nd Shift													/	/	/	
													/	/	/	
													/	/	/	
													/	/	/	
													/	/	/	
													/	/	/	

Comments, Additions, Etc.:

pH READING - PRIOR TO DUMPING TANKS. pH must be between 2.0 and 12.5 before dumping.

	TANK	DATE	°pH
	Stage 1		
	Stage 2		
	Stage 3		
	Stage 4		
	Stage 5		
	Stage 6		

Form-0403

WEEKLY MAINTENANCE CHECKLIST - (OVEN WASHER / OVEN LINE)

1. Remove bottom nozzles on stages not shaded.	Week	Pre-Wash	Stage 1	Stage 2	Stage 3	Stage 4	Stage 5	Stage 6
DOC.# _____ REVISION# _____ DATE: _____ AUTHORIZED BY: _____	Week 1							
	Week 2							
	Week 3							
	Week 4							
	Week 5							

2. Drain tanks (* Stages 2 and 3 only when instructed by supervision).	Week	Pre-Wash	Stage 1	Stage 2	Stage 3	Stage 4	Stage 5	Stage 6
	Week 1			*	*			
	Week 2			*	*			
	Week 3			*	*			
	Week 4			*	*			
	Week 5			*	*			

3. Blow out nozzles with water from the top down, wash out inside of tanks on stages not shaded. Hose down walls, risers, and floors in all stages ☐	Week	Pre-Wash	Stage 1	Stage 2	Stage 3	Stage 4	Stage 5	Stage 6
	Week 1							
	Week 2							
	Week 3							
	Week 4							
	Week 5							

4. Put bottom nozzles back on. Adjust nozzles in all stages as needed. Refill empty stages. Turn on pump and check each stage's pattern. Make any changes needed to ensure proper coverage.	Week	Pre-Wash	Stage 1	Stage 2	Stage 3	Stage 4	Stage 5	Stage 6
	Week 1							
	Week 2							
	Week 3							
	Week 4							
	Week 5							

5. Make sure halo is on. Set at (6-8 p.s.i.) ☐
6. Counter Flow Stage 6 (wide open) to Stage 5. Stage 5 at (6-8 p.s.i.) to Stage 4. Stage 4 at (3-5 p.s.i) to Stage ☐.
7. Clean all screens weekly. ☐
8. Wash drains (ditches) out in the oven washer and oven line areas weekly. ☐
9. Wash the sanitary tray (below the conveyor) out once a month on the Second Week. ☐
10. Wash back wall and floor (small room behind reinforce spray booth) out weekly. ☐
11. Check washer dryer for parts. ☐
12. Check enamel dryer for parts. ☐
* Recharge Stages 2 and 3 as directed by supervision.

WEEKLY SHUTDOWN STEPS
1. Take lights out and close doors on washer. ☐
2. Put fan back used for washer. ☐
3. Turn light off in reinforce booth. ☐
4. Put large cover and all grates back on ditche ☐
5. Cleaned furnace? (Yes / No) If you cleaned the furnace; Complete furnace checklis ☐
6. Check each stage to ensure: Tanks are full ☐   water's ☐   and drains are clo ☐.
7. Turn all fans off in oven line area ☐

Dryer

- Make sure blow-off hoses are positioned right.
- Make sure seams are dry.

Flow-coating

- Make sure to cover the entire part.
- Alert line leader to any changes that affect flow-coating. (Such as oil on parts, etc.)
- Stop line when needed if there is a change in the enamel that affects the flow-coating process.

Online Enamel Checks

- Pickup and Gravity checks performed every 30 minutes.
- Agitate enamel every 30 minutes.
- Wet Film Gauge checks between regular checks.
- Adjust pickup and gravity as needed to maintain parameters.

PORCELAIN SLIP RECORD CONTROL SHEET

OPERATOR & SHIFT: _____

OVEN LINE	TIME	TIME	TIME	TIME	TIME	TIME	TIME	TIME	TIME	TIME
DATE:										
TIME OF MEASUREMENT:										
SPECIFIC GRAVITY: 169-172										
PICK-UP: 55-60										
ADJUSTMENT MADE:										
OVEN LINE	TIME	TIME	TIME	TIME	TIME	TIME	TIME	TIME	TIME	TIME
DATE:										
TIME OF MEASUREMENT:										
SPECIFIC GRAVITY: 169-172										
PICK-UP: 55-60										
ADJUSTMENT MADE:										
OVEN LINE	TIME	TIME	TIME	TIME	TIME	TIME	TIME	TIME	TIME	TIME
DATE:										
TIME OF MEASUREMENT:										
SPECIFIC GRAVITY: 169-172										
PICK-UP: 55-60										
ADJUSTMENT MADE:										
OVEN LINE	TIME	TIME	TIME	TIME	TIME	TIME	TIME	TIME	TIME	TIME
DATE:										
TIME OF MEASUREMENT:										
SPECIFIC GRAVITY: 169-172										
PICK-UP: 55-60										
ADJUSTMENT MADE:										
COMMENTS:										

Pre-Dryer

- **Should be set to dry the enamel enough to set it, but not to completely dry it out**

Post Dryer

- Heat settings will vary with the type of parts. (We run at 300 degrees).
- We want the parts to be totally dry when they exit the post dryer.

Raw Ovens - Loading Log

DATE:

LOADER:

Part Number	Description	Ovens in Floor / Beg.	Ovens MadeToday	Ovens in Floor / End	Total Raw Ovens, Etc. Hung
2208W121-61	Floppy/Electric (Long Tail)				
2208W122-61	Half/Gas (Solid Frt. / Big Hole-Top)				
2208W123-61	Double/Gas (Solid Frt. / Big Hole-Top)				
2208W147-61	Large RV				
2208W148-61	Small RV				
2208W149-61	Floppy/Gas (Short Tail)				
2210W098-61	Half/Elec (Front w/slot)				
2210W101-61	Double/Elec (Front w/slot)				
2210W117-61	Jenn-Air 27"				
2210W125-61	Half/Elec (No slot / Small Hole-Top)				
2210W126-61	Doub/Elec (No slot / Small Hole-Top)				
2210W145-61	Jenn-Air 30"				
2210W146-61	Jenn-Air 27" (New)				
2210W152-61	Brand's 27"				
2210W153-61	Brand's 30"				
3401F016-61	Broiler Pan				
3407W005-61	Broiler Drawer				
3413F007-61	Broiler Insert				
3604F257-61	Broiler Shield				
4009F087-61	Pan Liner				
4011F243-61	Burner Box Bottom				

FORM 0305

Respray Log										
Oven Number Description	Black Lines	Burn Off	Blow-Out or Grease	Drain Lines	Handling	Heavy	Lumps	Outgass	Scars	Other
2208W121-19 Floppy / Elect.										
2208W122-19 Half / Gas										
2208W123-19 Double / Gas										
2208W147-19 Large RV										
2208W148-19 Small RV										
2208W149-19 Floppy / Gas										
2210W098-19 Half / Elect.										
2210W101-19 Double / Elect.										
2210W117-19 (J/A) 27"										
2210W125-19 Half / Elect.										
2210W126-19 Double / Elect.										
2210W145-19 (J/A) 30"										
2210W146-61 (J/A) 27" New										
2210W152-19 Brands 27"										
2210W153-61 Brands 30"										
3401F016-19 Broiler Pan										
3407W005-19 Broiler Drawer										
3413F007-19 Broiler Insert										
3604F257-61 Broiler Shield										

CHECKLIST FOR FABRICATION

Larry L. Steele
Mapes & Sprowl Steel, Ltd.
Elk Grove Village, IL

Abstract
This paper is a presentation of ways to achieve cost savings by the use of proper techniques in the manufacture of parts intended to be porcelain enameled. Areas from receiving the steel up to and including the finishing system are discussed. Incoming material properties, press operations, die/tooling maintenance, forming/drawing lubricants, welding/fastening, and different types of process soils and their effects are all discussed along with the relative impact of each of these process parameters.

In the manufacture of parts, primarily from sheet steels, there are many things that must be considered in order for this process to be successful. The material used to make parts has specific properties; the press operation(s) have variables; operation of the tooling used to make parts must be understood; lubricants will play a very important role; the process joining of metal parts should be understood in order to be successful; and, soils and contaminants that will be added to the part must be identified and addressed to insure consistently good finishing of the fabricated part.

Know Your Material
Whatever steel is used to make parts has specific properties that will have an impact on the success of the forming process. It is important that you know which properties affect the manufacture of each part. If your part uses drawing, the r-value, or measure of the materials resistance to thinning, will be of importance. If your part uses stretch, the n-value will provide a good measure, as will the percent total elongation. If your part requires simple bending, perhaps the yield strength will be the primary property of importance. Depending upon the severity of the forming process, surface roughness may also be of importance. Some steels are produced to provide a rougher surface that will assist in carrying more forming lubricant into the die. Typical variations in surface roughness will have, virtually no impact on fired porcelain enamel appearance, but may be detrimental to the final appearance finish on painted and/or chrome plated parts. However, very few parts will use only one of these parameters. Therefore, it is important that those parameters that will have the greatest effect on the successful forming your part be known.

It is also wise to understand that properties of steel, as delivered to you, will vary. There is very little, if any, property variation within the body of today's continuously cast steel products. However, these products will have some degree of variability between coils of steel. There are, typically, well defined limits of this variation. Your vendor of the steel products can provide you with information relative to the particular product he is supplying. As long as you have worked with this supplier in the development of a particular part prior to actual production, very few problems should arise as a result of the steel.

Know Press Operation
Once you understand the material you are using to produce a part, it is imperative that you fully understand the process of actually making that part. Damage control cannot be emphasized enough. Every blank, or coil end, that you save through damage control is, in reality, free steel.

When material is handled to bring it to the press operation, care must be taken to insure that it is not damaged. Stacking of skids in your inventory can cause damage to the top sheets if proper care is not taken. Blanks must be protected to insure damage free storage-heavy paper covering, band and clip protectors, no double stacking unless absolutely necessary. This will result in a scrap blank. Although it is not recommended that you use unoiled steel, the use of this type product significantly increases the importance of damage control. Likewise, handling of steel coils is important to insure no damage. Coil stock should not be double stacked. If absolutely necessary, adequate protection must be provided between coils. Coils should *never* be set directly on any floor. It is amazing just how much damage a small stone or pebble or some other particle can do should a coil be set down on it. A dent can be formed that will go in for several laps. The dent will very often result in a scrapped part or blank. This is material that you have paid for and that will end up in your scrap bin.

When using coils, it is important to understand operation of the coil reel and the flattener. The brake on the reel should be properly adjusted to insure smooth flow of steel from the coil. Jerking can create scratches or gouges of the surface before the material gets to the press. This is extremely important should a product be ordered without mill oil. It is recommended that some form of protection be used between the mandrel sectors and the ID of the coil. As steel is consumed, without proper protection, creases can form in the coil – due to mandrel sectors – with a significant amount of steel remaining (perhaps as much as 4" of sidewall). On appearance critical parts, this can result in scrap parts after all the costs of fabrication have been added.

While very few press lines have small roll levelers, many have large roll flatteners. It is imperative to understand that these can only remove coil set. Depending on how they are set, they can create upkick, downkick, or flat strip. The particular tooling requirements will determine this parameter. Should your coil stock have wavy edge, or non-edge buckles, a flattener will not correct this condition. However, it is possible to induce a wavy edge condition with an improperly adjusted flattener, or one that is in need of maintenance. Feed rolls – feeding the strip into the press – are also capable of inducing a wavy edge condition. Most of the rolls in these units, when originally manufactured, are made with a center crown to assist in material tracking. Over time, this will wear. These rolls can, actually, become dished out. Should this happen, it is possible to impart enough pressure with the adjusters that there is actually a small amount of cold work being done to the strip edge. When this occurs, a wavy condition will result. The answer to this problem is a good understanding of the unit's operation and proper maintenance.

Before you form your part, it is important to lubricate to reduce friction. As the part is being set up, it is imperative that the lubers are in good working order. If the luber is a roller applicator, the same condition described for feed rolls and flattener rolls – dished rolls – will result in poor application of the lube. Worn rolls will make it almost impossible to insure consistent lubricant application and will also result in too much lubricant being applied while trying to insure adequate lubrication over the entire part. The latter results in wasted lubricant. Again adding to the cost to produce your part. Should you have a spray lube system, it is imperative that the lube heads be aimed properly. The draw lube will be of no benefit if it is sprayed into the air instead of on the part surface.

As very few parts are formed in one stage, if there is a transfer system, it is very important to insure proper operation of the transfer equipment. This might be transfer fingers, suction cups, conveyor belts between presses, or robotic transfer. Whenever a part is handled, there is opportunity for damage. Damage results in lost parts and increased cost per good part. If part transfer systems are employed in the manufacture of a particular part, it is imperative that proper maintenance be performed. Should there be problems, proper training of the operator will help insure the bad conditions are observed and corrected without significant part loss.

Die/Tooling Maintenance

As parts are being formed, it is imperative that the operator be adequately trained in the operation of the form/draw dies. Many dies have draw beads to control metal flow into the die. As metal flows over these beads, a small amount is worn off the surface of the steel. This is usually in the form of a powdery type substance. Over time, this will build up on the draw beads and create scratches in this area of the part. As this "build-up" continues, it can reach the point where metal flow is impeded to the point that breakage will occur. Likewise, as metal particles continue to build up in a die, scratches can occur from such build-up on approach radii. This will result in the need to metal finish in order to maintain consistently good part surface. Operating parameters must be defined such that, as build-up occurs, the point where it is no longer acceptable is known. Once this point is reached, the operator must have the authority to stop production until the situation is corrected. Continued operation beyond this point will result in additional, *and unnecessary*, costs of manufacturing the part. To remove pick-up from draw beads and other die surfaces, the area should be honed. As most tools are hardened, in some manner, a grinder should NEVER be used on the die surfaces. This will only result in an unhardened area of the tooling and, eventually, will create the need to rework tools long before they should have been. Again, this will increase the cost to manufacture a part.

Hold Down/Pad Pressures must be properly maintained. Pressures too low will result in wrinkling. Wrinkling will result in areas of build-up. Too high of hold down pressures can result in excessive frictional forces on the steel. This will lead to build-up problems sooner that normal and could lead to part breakage. There are parameters that are operator controllable and must be well defined. Limits should be set so that, if exceeded, the operator will take appropriate corrective action before nonconforming parts are manufactured.

Be cognizant of the amount of shear burr in areas of trimming or punching. As tooling wears, burr will increase. Burrs are thin, highly cold worked areas of the steel. They are prone to breaking off and adding to build-up and/or scratching problems. Both of these will result in increased amounts of required metal finishing in order to produce good parts. Also, by sharpening punches and trim sections of tooling more often, the life of that particular part of the tooling is extended. It is a good idea to make tooling in such a manner that the trim sections are separate from the rest of the tooling. In this way, the trim section can be replaced when dull, possible without the need to physically remove the die from the press. This will result in increased up time and lower costs per part.

Draw Lube
In the forming of any part, proper lubrication is a must. Oils provided by the mills are primarily mineral oil with a small amount of rust preventative added. Mineral oil is ***not*** a good forming lubricant. Lubricant application was discussed under the "Press Operation" section. However, it is very important that there be consistency in lubrication. The lube supplier will assist in establishing operating/control parameters for the particular lubricant used to make your part. Make sure that the proper individuals know these parameters and adhere to them! Too often, individual operators will come up with some "Foo-Foo Juice" concoction that they will apply to address some problem. As this is, most often, an uncontrollable parameter, it is strongly recommended that this type practice be prohibited. One way to insure consistency in forming lubricants is to purchase premixed lubricants. These are usually a little more expensive, but each individual forming operation will dictate if this approach can be done in a cost effective manner. Your lubricant supplier will work with you in this arena. If there are forming problems, adequate quality control procedures should be in place to address. As previously discussed, insure that the lubers are working properly and applying the lubricant to the correct areas. Forming/draw lubricants reduce friction between the steel and the die surfaces and help to control metal flow into the form/draw die. Lubrication should be applied to each part/blank at the beginning of the forming process. Do Not apply intermittently or "*when needed*"! Inconsistent lubricant application will result in increased build-up issues in the forming dies. As has been stated before, this will only serve to increase the amount of nonconforming parts and the cost to produce good parts.

Welding & Fastening
Operators should be familiar with the joining process employed in the manufacture of your part. Spot welds are used quite often in the manufacture of parts. A good spot weld will never produce sparks when being made. If sparks are observed coming out of the welded, it could be due to misalignment of either the tips or the pieces being welded, improper settings of the welder, or contaminants in the weld area. Proper maintenance of weld tips is mandatory. Specific operating parameters must be established – and adhered to! As most weld programs are computer controlled, today, there should not be need for much, if any, adjustment unless something in the process has changed. Should frequent adjustment of the weld cycle program be required, the true root cause for this must be determined and addressed. Contamination of steel surfaces can be a contributor to variability of material coming into the weld area. Mill oils typically do not create problems in resistance welding. However, mill oils, in combination with draw lubes and other contaminants from press operations, could create variable conditions coming into the weld area. Likewise, organic residue in the area of the weld could leave carbonaceous deposit after welding that might be difficult to remove and contribute to gassing problems in porcelain enamel. While it is not always possible, it is desirable to clean parts prior to the welding operation. However, it is important that parts be welded very soon after cleaning

so as to avoid the formation of rust. This will significantly reduce, or eliminate, contamination issues.

Likewise, fusion welds – TIG & MIG – are, typically not recommended for parts intended for porcelain enameling. Because the steel is taken back to the molten state and allowed to resolidify, there is the probability that the conditions employed by the steel producer to eliminate porcelain enamel fishscale will be negated and fishscale will result in the weld area, and possible the heat affected zone, after the enameling operation.

Mechanical fastening/joining of parts is gaining in popularity. In this process, the two pieces of steel are, typically, drawn into a cup shape and then the metals are squeezed to "mushroom" the cup section. This results in a "button" that is of greater diameter than the drawn section. In this manner, the two pieces are permanently joined. Some manufacturers are currently using this technique for parts that are porcelain enameled. These joints do not seem to be negatively affected by the high temperatures associated with the porcelain enamel firing process. Reportedly, this type system requires less maintenance than a "typical" spot welding system.
While spot welds, when properly performed can be used on exposed porcelain enamel areas, it is unlikely that mechanically fastened parts would be practical in exposed applications for porcelain enameled parts.

Soils
Know ALL Your Process Soils! It is imperative that sources of all process soils be known. Mill oil, draw lubes, grease, hydraulic fluid, rubber compounds, floor dirt, plastic, and others are all potential soils on parts. Mill oils are usually very light in application and do not present significant problems. However, when in contact with some of the other contaminants, chemical reactions can occur that create totally new compounds. Grease and hydraulic fluids from the press operation should be kept off parts. When these begin to appear (press operator should be cognizant of this), corrective action should be taken immediately. Rubber/compound feed rolls will eventually deteriorate. When this happens, the rolls could leave particles on parts – intermittently – creating another potential source of contaminants. Operators must be aware of these conditions, as tracing intermittent problems is extremely difficult. As was previously stated, coil and blank stock must be kept off the shop floor. If the typical shop floor dirt is allowed to get on the steel surface, it has the potential of contaminating much of the forming/fabrication process. When this occurs, it is very difficult to determine root cause and the potential interactions with other, "normal" soils are unpredictable. When parts are transported in carriers between stages of fabrication, there is the possibility that the carriers might contribute a source of contamination. As wear and deterioration is observed on carriers and totes, it should be correct immediately.

Cleaning and Finishing
Know ALL Your Process Soils! This point cannot be made strongly enough. It is imperative that all sources of process contamination be identified. If the sources of soils cannot be eliminated, then they must be controlled. Determine if there are compatibility issues with the known soils and the forming/draw lubricant. Work with your lube supplier to determine any

potential issues. What happens when the part with process soils and forming lubricant gets to the cleaner stage? How much time is there between the last fabrication operation and the cleaner stage? What happens to lubes and process soils over time? What effects do the answers to these questions have on the effectiveness of the cleaner? Work with your cleaner supplier to address these potential issues. Have the lube supplier and cleaner supplier work together to address issues in cleaning. If possible, have the same chemical supplier be the source for both your draw lubes and your cleaners. This is the most desirable arrangement. As long as all process soils have been identified and are in control, there should not be any issues coming out of the cleaner stage.

Regardless of the finishing operation – porcelain enamel, paint, plating, etc. – specific, controllable operating parameters must be developed for the finishing system. All operating personnel must adhere to these controls.

Conclusion
It is hoped that the ideas, thoughts, and suggestions contained in this paper assist in the manufacture of parts and in the reduction of costs associated with the fabrication of parts that are intended to be porcelain enameled.

Environmental Issues

CONVERSION OF WET BOOTHS INTO DRY BOOTHS

Christophe Vanoplinus
Herding Filtration LLC

Abstract
Trends in the entire industry segment, from the application process to the production of process materials, all commonly have to deal with filtration issues. With environmental issues as well as dealing with permits, worker exposure, waste and containment. Herding reviewed a number of case histories where Herding's unique media has been used to solve a multitude of these issues by either retrofitting clients existing plants or integrating the media into new booth and collection designs to address the issues.

The porcelain industry, like other processing industries have also been challenged to reduce the large costs associated with energy and chemical or water usage. Trending towards the use of improved dry filtration solutions rather than expensive scrubber and wet booth technology, Herding's media can handle a wide range of collected streams at various load levels, and still fulfill an industry best clean exhaust emission result between 0.00004 and 0 .0004 grains/cubic foot.

The following slides are a partial representation from the slides used during the full presentation.

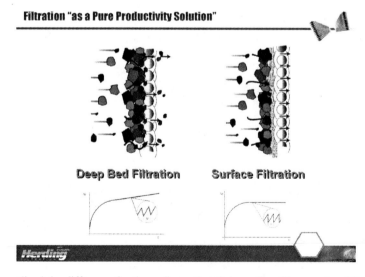

Herding outlined the difference for the audience that their media utilizes surface filtration, not media that can clog which would result in reduced air volume levels once a media becomes loaded.

Filtration "as a Pure Productivity Solution"

Rigid elements with filter coating
Complete surface filtration

Results in:

1. **Long lifetime** (permanent media)
2. **High efficiency**
   (example : SiO2-concentration lower
   0.03mg/m3)

Other advantages:
Low emissions : in gr/ft3 or mg/m3 (not in %) : No issues with permits
No downtime : No production loss
Long lifetime : No landfill - No filter change outs
Product Recovery – no contamination

Herding outlined the construction of the media, and its long life guarantees.

The Right Technology has to Solve Increasing Difficult Issues

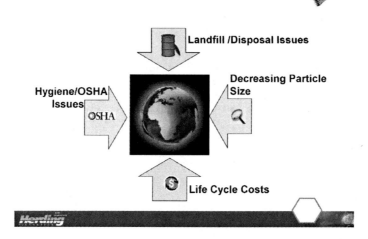

Landfill /Disposal Issues

Hygiene/OSHA Issues — OSHA

Decreasing Particle Size

Life Cycle Costs

The message was certainly made that engineers and plant management have a multitude of issues to deal with today, and in the future.

Grinding & Polishing booth – mfg ceramic bowls

Robotic & Manual Booth
With insert technology
Fanuc as result of
partnership

One case history was the full integration of the Herding media into a new modular booth and integrated collector design (Patent Pending) using Herding's Automotive expertise, and known downdraft booth concepts to optimize the process dust collection, containment, and recovery.

Conversion wet coating & finishing systems to dry filtration

Another case history was a client's removal of old traditional cartridge technology which was being replaced every 6 months, and could not meet the new emission performance levels. At the time of the Porcelain Technical Forum the unit was still operating after 18 months without any service disruption.

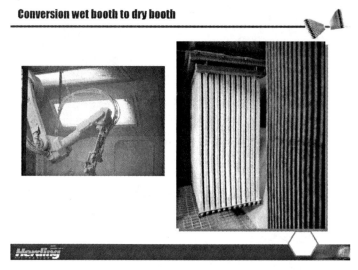

Conversion wet booth to dry booth

Another major point made in the presentation was the media's ability to survive in glazing, grinding, or spraying applications regardless of the streams moisture or abrasive nature. Note the particular use of a booth design where the media is directly located inside the robot booth. (Running for 5 years without change out of filter media)

Conversion wet booth to dry booth

Herding Filter Unit with open door

-fine SiO$_2$ dust

-after 10 years in operation

-with same filter elements

Given the industries need for uptime, and productivity for keeping plants profitable, the example shown above to summarize the presentation was even extremely fine dust, abrasive in nature and sticky, can still be processed in Herding units for the normal range of company guarantees of 7-10 years.

Other advantages include smaller footprints and floor space advantages, energy savings and compressed air costs. In addition, re-circulated air (given the extremely clean) exhaust can yield plant heating and air conditioning savings.

CURRENT REGULATORY ISSUES

Jack Waggener
URS Corporation

Abstract
The Occupational Safety and Health Agency (OSHA) is close to finalizing rules on hexavalent chrome (Cr^{+6}) in industrial processes. Dramatically lower limits are proposed by a factor of 10 to be reached by the year 2010. The Environmental Protection Agency (EPA) is also evaluating waste water discharges. The Toxic Release Inventory Report (TRI) ranked the porcelain enamel industry No. 13 on their list, however errors were noted by industry specialists and this has been addressed by the EPA regarding the Toxic Weighted Pound Equivalent Pollutants (TWPE). An example used was the use of sodium nitrite in enameling operations. Reasonable limits regarding the porcelain enamel industry are expected following further contact with the EPA later in 2006.

Low PEL
Impacts Many Processes in P.E. Plants &
Suppliers

Cr Oxides (PE)
Welding SS
Cr Sealers (Paint Lines)
Cr Electroplating
Welding Mild Steel
Chromates
Polishing/Grinding
Anodizing

4

Regulatory Schedule

Proposed	**Oct. 4, 2004**
Comments	**Jan. 3, 2005**
Public Hearing	**Feb. 1, 2005**
– 12 days/longest ever	
– Industry, Unions, OSHA, NIOSH	
Final Rule	**Feb. 28, 2006**
Initial Monitoring	
& Compliance	
More than 20 Employees	**Nov. 27, 2006**
Less than 20 Employees	**May 30, 2007**

5

PEL
Health Based for Lung Cancer

OSHA "1.0 ug/m³
(based on conservative model)
Industry 10 to 25 ug/m³
(model plus no anecdotal facts)
Unions "less than 0.25 ug/m³"
Final: 5 ug/m³

6

Cr⁺⁶ PEL
Occupational Exposure Limits:
Comparison of Selected Countries
(Trading Partners)

Country	Occupational Exposure Limit
United States	
— OSHA Final	5.0 ug/m3
— OSHA Current	52 ug/m3
Japan	50 ug/m3
European Union	50 ug/m3
France, Germany, UK, Finland	50 ug/m3
China	50 ug/m3
India	50 ug/m3
Sweden	*20 ug/m3*
Denmark	*5 ug/m3*

7

PEL Needs to be Technically Feasible

1) **Difficult to Measure (OSHA Method 215)**
   (Monitoring: Initial Periodic)
2) **Engineering Controls (Latest: May 31, 2010)**
   Local Ventilation, Fume Supp.
3) **Substitutes**
4) **Lastly, respirators**
   (could be air supplied)

8

Waste Water

EPA Proposes (8/29/05)

To Evaluate

Revising the

Porcelain Enamel Effluent Limitation Guideline (1982)

9

Driver

By CWA

    Required to re-evaluate ELGs

    every 2 years

In 1980s & 1990s

    EPA did not do

Note:

    **2003, Metal Products & Machinery ELG
(MP&M)**

No Rule

10

Schedule

Aug. 29, 2005 Federal Register

    Notice of "Preliminary"

        2006 Effluent Guidelines Plan

Oct. 28, 2005, Comments (EPA Errors)

March, 2006, PEI meets w/ EPA

September, 2006, Final Plan

What will EPA do w/PE?

11

EPA evaluated:

> **56 Industry Categories**
>
> **450 Industry Subcategories**

Based on Estimated TWPE

> **"Toxic Weighted Pound Equivalent Pollutants"**

From

Direct Discharge Reports (PCS)

&

Toxic Release Inventory Reports (TRI)

PE Ranked #13

12

Impacted Industries

EPA Evaluates:	(1000) TWPE
1. Pulp & Paper	4,600
2. Steam Electric Power	2,400
3. Organic Chemicals	2,300
4. Petroleum Refining	670
5. Pesticides	610
6. Non-Ferrous Metals	510
7. Ore Mining	470
8. Inorganic Chemicals	420
9. Ruber Manuf.	180
10. Textiles	160
11. Fertilizers	150
12. Plastic Molding	98

13. Porcelain Enamel   92

13

EPA
Mistakes & Issues

Includes Many Non PE Plants

Bad Evaluation of Pollutants

Included Storm Water

Included Metal Finishing (MF)
Most Flows are MF

Increased Toxic Factors

14

14 PE Plants Evaluated

American Standard, Salem, OH	TRI
Electrolux, Springfield, TN	TRI
Hanson Porcelain, Lynchburg, VA	TRI
Kohler, Cast Iron, WI	TRI
Maytag, Newton, IA	TRI
Maytag #1 & #3, Cleveland, TN	TRI
Roper, Lafayette, GA	TRI
VITCO, Nappanee, IN	TRI
Whirlpool, Clyde, OH	TRI
Whirlpool, Tulsa, OK	TRI
State Inds., Ashland City, TN	TRI & PCS
Briggs, Knoxville, TN	PCS

15

17 NON PE Plants

Electrolux, Webster City & Jefferson (IA)

GE, Louisville (KY), Decatur (AL), Bloomington (IN)

Maytag, Searcy (AR), Amana (IA), Herrin (IL)

Whirlpool, Evansville (IN), Ft. Smith (AR),
**   Findlay (OH), Marion (OH)**

Kohler, Searcy (AR)

W.C. Wood, Ottawa (OH)

Eljer, Salem (OH) (Closed)

American Trim, Wapakoneta (OH)

16

Comments Refute TWPE

Based on TRI

EPA	PEI
88,749	<231
	(<0.3% of EPA)

Example:

   Sodium Nitrite

       26,000 # Purchased (not discharged)

       1,600 # related to PE

          1,200 # final rinse

          400 # PE mill addition

       24,400 # in MF

Reality: Nitrite rapidly oxides to Nitrate (1000 times less
**   toxic)**

17

Comments Refute TWPE (Con't.)

Based on Direct Discharge Reports (PCS)

EPA	PEI
3,478	<17
	(<0.5% of EPA)

18

Follow-up

Oct. 28, 2005
 PEI Comments Submitted

March, 2006.
 Follow-up Meetings/Discussion w/EPA

September, 2006
 Final Decision from EPA
Looks Good
PEI is helping you!
 THANK YOU!

19

Author Index

68th Porcelain Enamel Institute Technical Forum